A-LEVEL
AND AS-LEVEL

I0657708

CHEMISTRY

Dr Michael C. Cox

Weymouth College
Learning for Life
This item is to be returned on or before
the last date stamped below

-6. JAN. 1997	- 1 MAR 2001	
-5. JUN 1997	- 3 APR 2001	
1L NOV 1997		
	- 5 DEC 2001	
20 APR. 1998	2 5 NOV 2002	
4. NOV. 1998	2 7 NOV 2014	
-9. FEB. 1999	2 8 SEP 2016	
1 1 MAR 1999		
2 7 APR 1999		
2 7 SEP 1999		

L

LONGMAN A AND AS-LEVEL REFERENCE GUIDES

Series editors: Geoff Black and Stuart Wall

TITLES AVAILABLE
Biology
Chemistry
English
Geography
Mathematics
Physics

Longman Group UK Limited,
Longman House, Burnt Mill, Harlow,
Essex CM20 2JE, England
and Associated Companies throughout the world.

© Longman Group UK Limited 1991

First published 1991

British Library Cataloguing in Publication Data

Cox, Michael
  Chemistry.
  1. Great Britain. Secondary schools. Curriculum
  subjects: Chemistry
  I. Title
  540.71241

  ISBN 0–582–06390–6

Designed and produced by
The Pen and Ink Book Company,
Huntingdon, Cambridgeshire.
Set in 10/12pt Century Old Style.

Printed in Singapore

ACKNOWLEDGEMENTS

I am especially grateful to the editors, Stuart Wall and Geoff Black, for their very helpful comments and guidance and to Professor Richard Kempa of the University of Keele for his extremely thought-provoking observations and constructive suggestions.

Above all, I wish to thank my wife, Maureen, for her continuing love, patience and support yet again.

HOW TO USE THIS BOOK

Throughout your A-level and AS-level course you will be coming across terms, ideas and definitions that are unfamiliar to you. The Longman Reference Guides provide a quick, easy-to-use source of information, fact and opinion. Each main term is listed alphabetically and, where appropriate, cross-referenced to related terms.

- Where a term or phrase appears in **different type** you can look up a separate entry under that heading elsewhere in the book.
- Where a term or phrase appears in **different type** and is set between two arrowhead symbols ◄　　　►, it is particularly recommended that you turn to the entry for that heading.

ACETYLATION

Acetylation is the **substitution** of the CH_3CO group into a **molecule** containing OH or NH_2 groups.

Acetylation, or **ethanoylation** to name it more systematically, is a specific example of an **acylation**. The usual 'acetylating' agents are ethanoyl chloride and ethanoic anhydride. They may be regarded as **electrophiles** undergoing nucleophilic attack by the molecules being 'acetylated': see Fig. A.1.

the first step could be considered as a nucleophilic addition:

$$R-O: \quad \overset{\delta-}{\underset{\delta-Cl \quad CH_3}{\overset{O}{\underset{\|}{C^{\delta+}}}}} \quad \rightarrow \quad R-\overset{+}{\underset{H}{O}}-\overset{O^-}{\underset{Cl}{\overset{|}{C}}}-CH_3$$

the second step could then be regarded as an elimination:

$$R-\overset{+}{\underset{H}{O}}-\overset{:O:^-}{\underset{Cl}{\overset{|}{C}}}-CH_3 \quad \rightarrow \quad \overset{O}{\underset{R-O \quad CH_3}{\overset{\|}{C}}} \quad + \quad HCl$$

the overall effect is a nucleophilic substitution of the H in ROH by the acyl group: $(CH_3C=O)$.

$$R-\overset{}{\underset{H}{O}}: \quad + \quad \overset{O}{\underset{Cl \quad CH_3}{\overset{\|}{C}}} \quad \rightarrow \quad \overset{O}{\underset{R-O \quad CH_3}{\overset{\|}{C}}} \quad + \quad HCl$$

Fig. A.1 Nucleophilic substitution with acyl chlorides

The **activation energy** of the first step may be quite low because the energy required for breaking only one of the bonds in the C=O group is provided by the simultaneous formation of the +O−C bond. An ionising solvent and **alkaline** conditions would facilitate the expulsion of the stable chloride ion and the **proton** in the second step which would also have a low activation energy. Hence the substitution would be fast.

ACID

An acid is a substance with a sour taste (Latin *acidus* – sour).

The Hon. Robert Boyle (1627–91), seventh son of the Earl of Cork, a founder of the Royal Society and of modern (scientific) chemistry, classified sharp-tasting substances as acids by their effect upon litmus, **carbonates** and **alkalis.**

Antoine Laurent Lavoisier (1743–94), the Frenchman famous for his explanation of combustion, invented the name oxygen because he believed the element was a constituent of all acids (Greek *oxys* – sour; *gennao* – I produce).

Sir Humphrey Davy (1790–1845), sometime President of the Royal Society, proved that 'muriatic acid' (now called hydrochloric acid) did not contain oxygen and Auguste Laurent (1807–53) then supposed that acids were compounds of hydrogen.

An acid is a substance that dissociates in water to form **hydrogen ions.** Svante Arrhenius (1859–1927), in his famous theory on the dissociation of electrolytes, proposed acids to be substances that split up in water to form hydrogen ions, $H^+(aq)$. Strong acids dissociate completely and weak acids dissociate partially. If $[H^+(aq)]$ and $[OH^-(aq)]$ represent the concentration of the **aqueous** hydrogen ions and **hydroxide ions,** then

$[H^+(aq)] > [OH^-(aq)]$ in acidic solutions
$[H^+(aq)] = [OH^-(aq)]$ in neutral solutions
$[H^+(aq)] < [OH^-(aq)]$ in alkaline solutions.

In 1923 Johannes Bronsted in Denmark and Thomas Lowry in England independently proposed a more general definition of acids and bases now known as the **Bronsted–Lowry theory**: an acid is a proton donor and a base is a proton acceptor. At the same time Gilbert N. Lewis in America advanced his more comprehensive electron-pair theory of a Lewis acid and Lewis base: an acid is an electron-pair acceptor and a base is an electron-pair donor.

The Arrhenius and Bronsted–Lowry theories can provide a quantitative measure of acidity (in terms of $[H^+(aq)]$ and pH) and acid strength (in terms of the **acid dissociation constant**). The Lewis theory can provide a qualitative interpretation of **complex** formation and of organic **reaction mechanisms** in terms of **ligands, electrophiles, nucleophiles** and the formation of **coordinate** (dative) **bonds.**

ACID AMIDE

Amides are considered as derivatives of **carboxylic acids** because their formulae contain the **functional group** $-C-NH_2$.

$$\overset{\|}{\underset{O}{}}$$

Amides can be made by reacting **acyl chlorides** with ammonia or by **refluxing** an anhydrous mixture of a carboxylic acid and its ammonium salt for several hours, for example:

$$NH_3 + CH_3\overset{|}{\underset{|}{C}}=O \rightarrow CH_3-\overset{|}{\underset{|}{C}}=O + HCl$$
$$\qquad\quad Cl \qquad\qquad NH_2$$

$$CH_3CO_2^-NH_4^+ \xrightarrow{\;-H_2O\;} CH_3CONH_2$$

NOMENCLATURE

The systematic names of the **aliphatic** amides include the C of the C=O group in the carbon chain: for instance CH_3CONH_2 ethanamide (traditionally called acetamide). If one or both H-atoms in the $-NH_2$ are replaced by alkyl groups the N is included in the name: hence $CH_3CONHCH_3$ is called N-methylethanamide. The simplest diamide, $CO(NH_2)_2$ is **urea** (or carbamide) which may be seen as a derivative of carbonic acid, $CO(OH)_2$.

PHYSICAL PROPERTIES

Amides (except for methanamide – m.p. 2.5°C) are white crystalline solids. Amide molecules hydrogen—bond together but not predominantly as **dimers**; consequently, their **melting points** are usually higher than their parent carboxylic acids; for example ethanamide (**molar mass** 59 g mol⁻¹; m.p. 82.2°C) and ethanoic acid (molar mass 60 g mol⁻¹; m.p. 16.5°C). **Hydrogen bonding** makes the lower amides very soluble in water as well as in organic liquids.

Ethanamide smells of 'dead mice' and is now listed as a carcinogenic compound.

CHEMICAL PROPERTIES

lone pair of electrons on the nitrogen atom becomes part of a delocalised system with the π-electrons of the carbonyl group

Fig. A.2 Electron delocalisation in amides

Amides are far less reactive and far less basic than **amines** because the N—atoms's **lone pair** of electrons are delocalized with the π-electrons of the C=O bond (see Fig. A.2). Amides can be converted into amines by powerful **reducing agents**:

$$R-\overset{\|}{\underset{O}{C}}-NH_2 \xrightarrow[\text{followed by } H_2O \text{ for hydrolysis}]{\text{LiAlH}_4 \text{ in dry ethoxyethane}} R-CH_2-NH_2$$

or by bromine under **alkaline** conditions in the **Hofmann degradation** reaction. Powerful dehydrating agents convert amides into **nitriles**:

$$\underset{\underset{O}{\parallel}}{R-C-NH_2} \xrightarrow{\text{Heat with phosphorus(V) oxide}} R-CN$$

Nitrous acid replaces the $-NH_2$ groups in amides by $-OH$ groups and liberates almost quantitatively one mole of N_2 gas for every one mole of $-NH_2$ group:

$$\underset{\underset{O}{\parallel}}{R-C-NH_2} + HNO_2 \xrightarrow[\text{warm}]{\text{NaNO}_2/\text{HCl(aq)}} N_2(g) + \underset{\underset{O}{\parallel}}{R-C-OH} \quad \text{carboxylic acid}$$

Amides are hydrolysed by refluxing with aqueous strong acids or alkalis:

$$\underset{\underset{O}{\parallel}}{R-C-NH_2} + H_2O \xrightarrow[\text{reflux}]{H^+ \text{ or } OH^-} NH_3 + \underset{\underset{O}{\parallel}}{R-C-OH} \quad \text{carboxylic acid}$$

ACID ANHYDRIDE

Inorganic acid anhydrides are **oxides** of non-metals that react with water to form an **acid**.

CO_2, SO_2 and SO_3 are the anhydrides of carbonic (H_2CO_3), sulphurous (H_2SO_3) and sulphuric (H_2SO_4) acid, respectively. Some oxides are mixed anhydrides because they produce two acids when they react with water; for example, N_2O_4 forms a mixture of HNO_2 and HNO_3 with water.

Organic acid anhydrides are colourless, sharp-smelling liquids regarded as derived from two molecules of **carboxylic acid** by the removal of a water molecule; for instance:

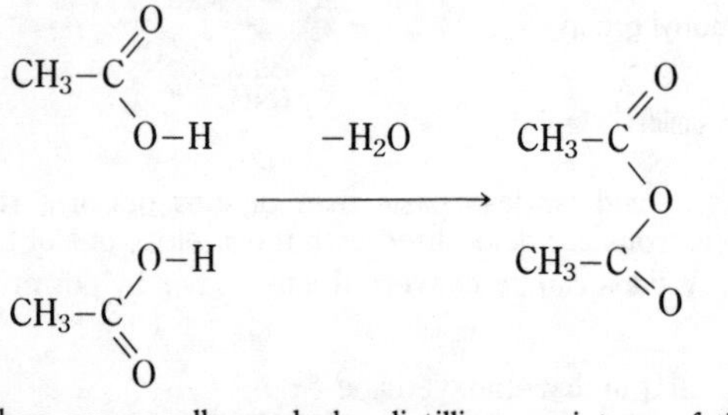

They are usually made by distilling a mixture of the **acyl chloride** and the sodium salt of the carboxylic acid in a dry apparatus and collecting the anhydride as the distillate in a dry receiver:

$$CH_3CO_2^-Na^+(s) + \underset{\underset{Cl}{|}}{CH_3C{=}O(l)} \xrightarrow{\text{distil}} \underset{\underset{O}{||}\;\underset{O}{||}}{CH_3{-}C{-}O{-}C{-}CH_3(l)} + Na^+Cl^-(s)$$

Acid anhydrides are less reactive than acyl chlorides and are used to prepare **amides** and **esters**, especially of **phenols** where direct **esterification** is not possible. For example, ethanoic anhydride can be used to prepare aspirin from 2-hydroxybenzoic (salicylic) acid (see Fig. A.3).

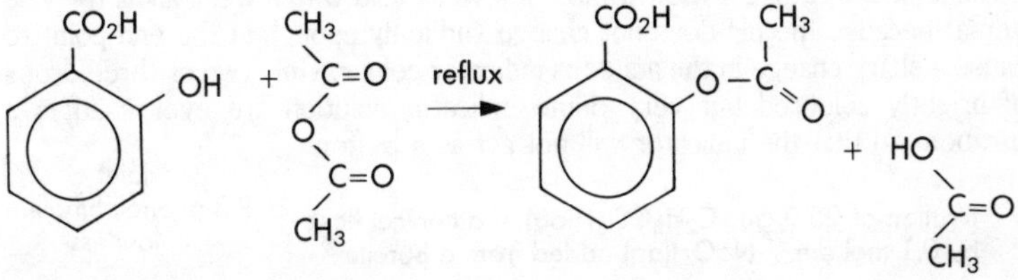

| 2-hydroxybenzoic (salicylic) acid | ethanoic (acetic) anhydride | (aspirin) 2-ethanoyloxy-benzoic acid | ethanoic acid |

Fig. A.3 Preparation of aspirin

ACID–BASE INDICATOR

An acid–base indicator is a substance whose **aqueous** solution changes colour with a change in pH. Acid—base indicators are coloured organic compounds whose molecular structure changes as a result of gaining or losing **protons**. They may be regarded as **weak acids** with the aqueous acid having a different colour from that of its aqueous **conjugate base**; see Fig. A.4.

Name	Weak acid form	pK_a	Conjugate base form
methyl orange	red	3.5	yellow
methyl red	red	5.1	yellow
bromothymol blue	yellow	7.0	blue
phenolphthalein	colourless	9.3	violet

Fig. A.4 Acid—base indicators

The colour of an aqueous acid—base indicator is governed by the ratio of the **concentrations** of its weak acid (HIn) and conjugate base (In$^-$), and this ratio is governed by the concentration of the $H_3O^+(aq)$ ion and the **dissociation constant** of the indicator K_{in} –

$$\frac{[H_3O^+(aq)][In^-(aq)]}{[HIn(aq)]} = K_{in}$$

When $[In^-(aq)] = [HIn(aq)]$, the acid–base indicator shows its mid-point colour and the pH of the solution will equal pK_{in}. And the indicator will change colour over a range of about 2 pH units. For example, for bromophenol blue, HIn(aq) makes the solution yellow, In⁻(aq) makes it blue and pK_{in} = 4.0. So aqueous bromophenol blue will be yellow if the pH is less than 3, blue if the pH is more than 5 and green when the pH = 4.0.

If an acid–base indicator is to be suitable for a titration, its pK_{in} should be equal to the pH at the end-point where pH changes most sharply. For example, phenolphthalein (pK_{in} = 9.3) would be a suitable indicator for the titration of aqueous propanoic acid by aqueous NaOH(aq) from a burette but methyl orange (pK_{in} = 3.5) would be most unsuitable: see Fig. A.5. There is no suitable indicator for the titration of a weak acid with a weak alkali (or vice versa) because the pH does not change suddenly enough at the end-point to cause a sharp change in the aqueous indicator colour. Only two or three drops of brightly coloured but very dilute indicator solution are ever used in a titration, so that the indicator will not act as a buffer.

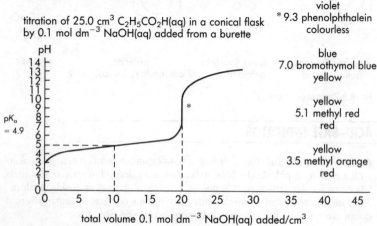

titration of 25.0 cm³ $C_2H_5CO_2H$(aq) in a conical flask by 0.1 mol dm⁻³ NaOH(aq) added from a burette

violet
* 9.3 phenolphthalein
colourless

blue
7.0 bromothymol blue
yellow

yellow
5.1 methyl red
red

yellow
3.5 methyl orange
red

Fig. A.5 pH-titration curve of aqueous propanoic acid

ACID–BASE TITRATION

An acid–base titration is a method of measuring with a burette the volume of aqueous acid equivalent to a measured volume of aqueous base and drops of acid–base indicator in a conical flask (see Fig. A.6).
◀ Titration, volumetric analysis ▶

ACID DISSOCIATION CONSTANT

The dissociation constant, K_a of an acid, HA, is the equilibrium constant for the reaction HA(aq) ⇌ H⁺(aq) + A⁻(aq) as given by the expression:

$$\frac{[H^+(aq)][A^-(aq)]}{[HA(aq)]} = K_a$$

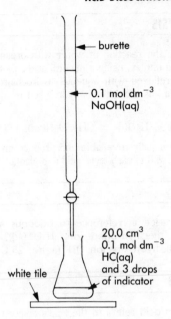

Fig. A.6 Acid—base titration

Acid	$=$	H^+	$+$	Conjugate base		$K_a/mol\ dm^{-3}$	pK_a
$H_3O^+(aq)$				$H_2O(l)$		1.0	0
$H_2SO_3(aq)$				$HSO_3^-(aq)$		1.5×10^{-2}	1.8
$HSO_4^-(aq)$				$SO_4^{2-}(aq)$		1.0×10^{-2}	2.0
$H_3PO_4(aq)$				$H_2PO_4^-(aq)$		7.9×10^{-3}	2.1
$HF(aq)$				$F^-(aq)$		5.6×10^{-4}	3.3
$HNO_2(aq)$				$NO_2^-(aq)$		4.7×10^{-4}	3.3
$CH_3CO_2H(aq)$				$CH_3CO_2^-(aq)$		1.7×10^{-5}	4.8
$H_2CO_3(aq)$				$HCO_3^-(aq)$		4.3×10^{-7}	6.4
$H_2PO_4^-(aq)$				$HPO_4^{2-}(aq)$		6.2×10^{-8}	7.2
$NH_4^+(aq)$				$NH_3(aq)$		5.6×10^{-10}	9.3
$C_6H_5OH(aq)$				$C_6H_5O^-(aq)$		1.3×10^{-10}	9.9
$HCO_3^-(aq)$				$CO_3^{2-}(aq)$		4.8×10^{-11}	10.3
$HPO_4^{2-}(aq)$				$PO_4^{3-}(aq)$		4.4×10^{-13}	12.4
$H_2O(l)$				$OH^-(aq)$		1.0×10^{-14}	14.0

INCREASING ACID STRENGTH →

INCREASING BASE STRENGTH ↓

Fig. A.7 Table of acid strengths

K_a measures the strength of an acid on a scale from 1×10^0 mol dm^{-3} (for the $H_3O^+(aq)$ ion) to 1×10^{-14} mol dm^{-3} (for the $H_2O(l)$ molecule). **Carboxylic acids** such as ethanoic acid are typically moderately **weak** acids with K_a values of around 1×10^{-5} mol dm^{-3}; see Fig. A.7.

◀ Equilibrium constant ▶

ACID HYDROLYSIS

Acid hydrolysis is the reaction of water with organic compounds, such as amides, esters and nitriles, using a mineral acid as a catalyst.

If an ester is refluxed with aqueous hydrochloric acid, the reverse of esterification occurs as the ester is hydrolysed to a carboxylic acid and an alcohol:

$$CH_3CO_2C_2H_5(l) + H_2O(l) \rightleftharpoons CH_3CO_2H(aq) + C_2H_5OH(aq)$$

The reactions are usually reversible and produce an equilibrium mixture of the original ester, water, carboxylic acid and alcohol.

ACIDIC

Acidic is a term used in reference to aqueous solutions in which the concentration of the hydrogen ion, $H_3O^+(aq)$, is greater than the concentration of the hydroxide ion, $OH^-(aq)$; at 25°C the pH is less than 7.

ACID STRENGTH

The strength of an acid refers to the acid's ability to lose a proton and is measured by the acid dissociation constant, K_a.

Strong acids such as hydrochloric acid are completely dissociated in water. Weak acids such as the aliphatic carboxylic acids are only partially dissociated in aqueous solution.

ACTINOID

The actinoids (actinides) are a horizontal row of elements from thorium to lawrencium in the f-block of the periodic table.

ACTIVATED COMPLEX

An activated complex is a short-lived unstable intermediate molecule or ion formed from the reactants and existing in transition state prior to the formation of the products.

In the S_N2 alkaline hydrolysis of 1-bromobutane, for example, the activated complex would be a species in which the attacking nucleophile (OH^-) and the leaving nucleophile (Br^-) are both partially bonded and part of the molecular ion in the transition state:

activated complex

ACTIVATION ENERGY

Activation energy, E_a, is a parameter in the **Arrhenius equation** for the dependence of the **rate constant**, k, upon temperature: $k = Ae^{-E_a/RT}$ where R is the **gas constant** and A is the Arrhenius (or pre-exponential) factor. The Arrhenius equation may be written as $\ln k = \ln A - E_a/RT$ and a value for the activation energy for a reaction may be determined from the gradient ($= -E_a/R$) of a graph of $\ln k$ against $1/T$; see Fig. A.8.

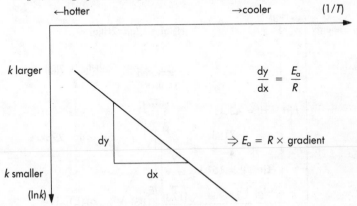

Fig. A.8 Determining the activation energy of a reaction

E_a and the energy barrier

The activation energy may be regarded as the minimum energy needed by the reacting particles (atoms, ions or molecules) to achieve the **transition state** so that a reaction may occur between them; see Fig. A.9.

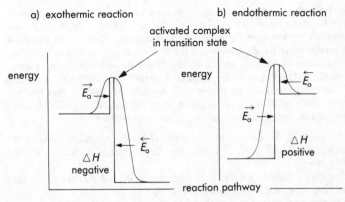

E_a = activation energy of forward ($\overrightarrow{E_a}$) and reverse ($\overleftarrow{E_a}$) reaction

Fig. A.9. Reaction profiles

Reaction profiles show the activation energies as barriers that must be surmounted before reactants can turn into products and before products can re-form into reactants. Sometimes the molecules in the transition state can be imagined to have formed an **activated complex** in which the old bonds have partly broken and the new bonds have partly formed; see Fig. A10. The activation energy could then be related to the difference between the partly broken and partly formed bond energies. The energy required to break strong **covalent bonds** may lead to high activation energies and explain the slow reactions of some organic molecules.

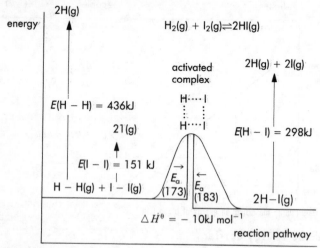

Fig. A.10 Formation of an activated complex in a transition state

E_a and the distribution of energy

In the Arrhenius equation, A may be related to the frequency of collisions between the reacting particles and $e^{-E_a/RT}$ to the fraction of collisions causing a reaction. A low value for E_a makes the fraction large (giving a fast reaction), whilst a high value for E_a makes the fraction small (giving a slow reaction). The distribution of energies can be shown by a **Maxwell–Boltzmann curve** – see Fig. A.11. The function $f(E, n)$ plotted on the y-axis makes the total area under the curve proportional to the total number of molecules and the area under any portion of the curve proportional to the number of molecules with energies in that range. Consequently, as the temperature rises

a) the peak of the curve moves to the right so the mean value of the function $f(E, n)$ (and therefore the mean energy of the molecules) increases,

b) the curve flattens so the total area under it (and therefore the total number of molecules) remains constant,

c) the area under the curve to the right of E_a (and therefore the number of molecules colliding with enough energy to cause a reaction) roughly doubles for every 10-degree rise in temperature.

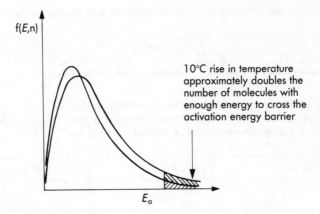

Fig. A.11 Maxwell–Boltzmann curve

So the higher the temperature the faster the reaction and the lower the activation energy the faster the reaction.

ACTIVE SITE

An active site is a place on the surface of a **heterogeneous catalyst** where reactants may be adsorbed and the reaction is catalysed. On an **enzyme** an active site is a place where the specific **substrate** becomes attached during the reaction catalysed by the enzyme.

The primary, secondary and tertiary structures of the enzyme determine the nature of the active site and specificity of the enzyme's catalytic action. Subtle changes in pH and in temperature may distort the tertiary structure of the enzyme sufficiently to destroy the active site and inhibit the catalytic activity of the enzyme.

ACYCLIC

◄ Aliphatic ►

ACYLATION

Acylation is the insertion of an acyl group into a molecule by reacting an **acyl chloride** or an organic **acid anhydride** with a compound containing an amino group or a **hydroxyl group**.

Acylation reactions may be used to esterify **phenols** because these compounds do not react directly with **carboxylic acids**. For example, phenyl benzoate can be prepared in the laboratory in good yield by shaking benzenecarbonyl (benzoyl) chloride with benzoic acid in aqueous alkali:

$$
\underset{\text{benzoyl chloride}}{C_6H_5\overset{\overset{\displaystyle O}{\|}}{C}-Cl} + \underset{\text{phenol}}{C_6H_5OH} \xrightarrow[\text{conditions}]{\text{alkaline}} \underset{\text{phenyl benzoate}}{C_6H_5\overset{\overset{\displaystyle O}{\|}}{C}-O-C_6H_5} + HCl
$$

The more important acylation reactions are listed in Fig. A.12.

Acylation is used to protect OH and NH_2 groups from, say, oxidation by nitric acid during the **nitration** of an **aromatic amine**; see Fig. A.13.

Reactant molecule	Product(s)	Class of compound
water H_2O	$R-\overset{\overset{\displaystyle O}{\|}}{C}-OH$	carboxylic acid
methanol CH_3OH	$R-\overset{\overset{\displaystyle O}{\|}}{C}-OCH_3$	methyl ester
phenol C_6H_5OH	$R-\overset{\overset{\displaystyle O}{\|}}{C}-OC_6H_5$	phenyl ester
ammonia NH_3	$R-\overset{\overset{\displaystyle O}{\|}}{C}-NH_2$	amide
methylamine CH_3NH_2	$R-\overset{\overset{\displaystyle O}{\|}}{C}-NHCH_3$	substituted amide
diethylamine $(C_2H_5)_2NH$	$R-\overset{\overset{\displaystyle O}{\|}}{C}-N(C_2H_5)_2$	substituted amide
$R-\overset{\overset{\displaystyle O}{\|}}{C}-O^-Na^+$	$R-\overset{\overset{\displaystyle O}{\|}}{C}-O-\overset{\overset{\displaystyle O}{\|}}{C}-R$	acid anhydride
benzene C_6H_6 (+ Friedel-Craft catalyst – $AlCl_3$)	$R-\overset{\overset{\displaystyle O}{\|}}{C}-C_6H_5$	phenyl ketone

Fig. A.12 Nucleophilic acyl substitutions (acylations)

Preparing 4-nitrophenylamine from phenylamine:

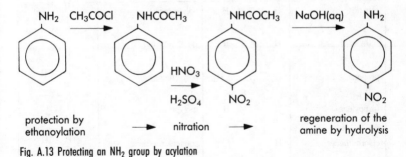

Fig. A.13 Protecting an NH_2 group by acylation

ACYL CHLORIDE

Acyl chlorides are extremely reactive **carboxylic acid** derivatives containing the **functional group** $-\underset{O}{\overset{\|}{C}}-Cl$.

Ethanoyl chloride (acetyl chloride) and benzenecarbonyl chloride (benzoyl chloride) are two important examples encounted in A-level chemistry. They are colourless, corrosive liquids that fume in moist air to form the corresponding organic acid and hydrogen chloride:

$$R-\underset{O}{\overset{\|}{C}}-Cl + H_2O \rightarrow R-\underset{O}{\overset{\|}{C}}-OH + HCl$$

The electronegative oxygen and chlorine atoms make the carbon atom a strongly electrophilic centre readily attacked by **nucleophiles** such as H_2O, ROH, NH_3, RNH_2. Acyl chlorides are reactive compounds that can be used to prepare **acid anhydrides**, **esters** and **amides** in very good yields.

ADDITION POLYMER

An addition **polymer** is a compound formed when a large number of small molecules combine to form a large **molecule**. The small molecules or **monomers** are usually **alkenes**. A homopolymer consists of identical monomers. A copolymer consists of different monomers. Addition polymers are **thermoplastic**. Some important poly(alkenes) are shown in Fig. A.14 overleaf.

ADDITION REACTION

An addition reaction is a reaction in which two molecules combine to form a single **molecule**. Electrophilic addition reactions are typified by those of propene $CH_3-CH=CH_2$ (see Fig. A.15 overleaf).

Monomer	Formula	Polymer	Structure
ethene (ethylene)	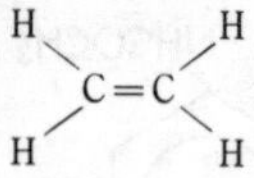 H₂C=CH₂	poly(ethene) (polyethylene)	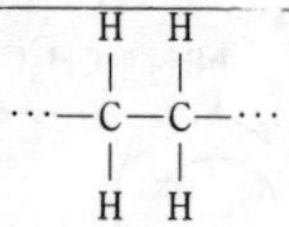 ···—C—C—··· (—CH₂—CH₂—)
propene (propylene)	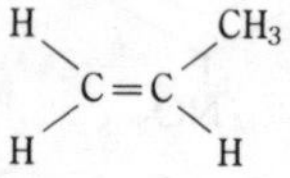 H, CH₃ C=C H, H	poly(ethene) (polypropylene)	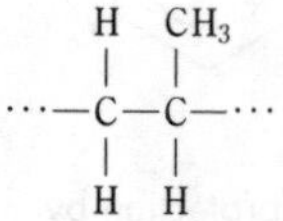 ···—C—C—··· (—CHCH₃—CH₂—)
chloroethene (vinylchloride)	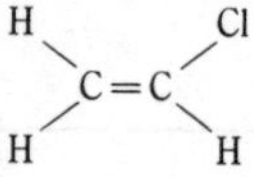 H, Cl C=C H, H	poly(chloroethene) (PolyVinylChloride)	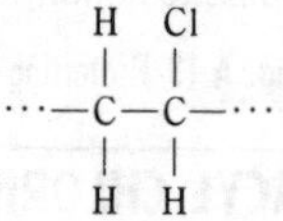 ···—C—C—··· (—CHCl—CH₂—)
tetrafluoroethene	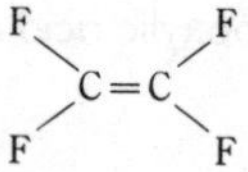 F, F C=C F, F	poly(tetrafluoroethene) (PTFE)	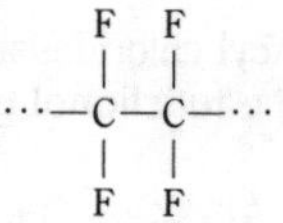 ···—C—C—··· (—CF₂—CF₂—)
phenylethene (styrene)	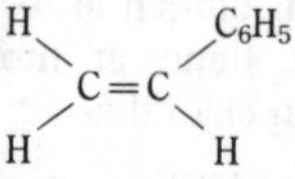 H, C₆H₅ C=C H, H	poly(phenylethene) (polystyrene)	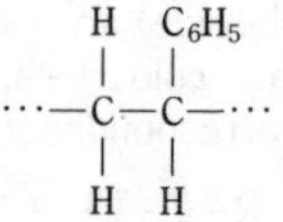 ···—C—C—··· (—CHC₆H₅—CH₂—)
propenitrile (acrylonitrile)	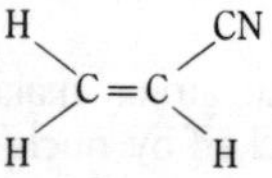 H, CN C=C H, H	poly(propenitrile) (polyacrylonitrile)	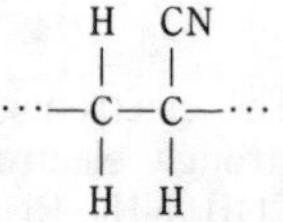 ···—C—C—··· (—CHCN—CH₂—)
2-methylpropenoate (methyl methacrylate)	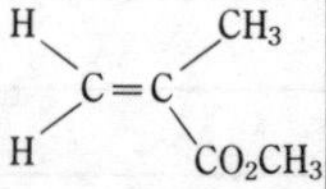 H, CH₃ C=C H, CO₂CH₃	poly(2-methylpropenoate) (polymethylmethacrylate)	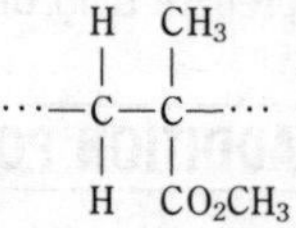 ···—C—C—··· (—CH₂—C(CH₃)(CO₂CH₃)—)

Fig. A.14

Molecule	Reagent and conditions	Main product(s)
Br—Br	bromine as a vapour or in an organic solvent	CH_3—$CHBr$—CH_2Br 1,2-dibromopropane
Br—OH	bromine in aqueous solution	CH_3—$CHOH$—CH_2Br 1-bromopropan-2-ol
H—Br	hydrogen bromide gas or conc. hydrobromic acid	CH_3—$CHBr$—CH_3 2-bromopropane
H_2O	conc. sulphuric acid followed by water	CH_3—$CHOH$—CH_3 propan-2-ol

Fig. A.15 Electrophilic addition reactions of propene

Alkenes undergo electrophilic addition reactions because the high π-electron density makes the C=C bond act as a nucleophilic centre. Electrophilic additions occur across the **double bond** and proceed by a **mechanism** illustrated by the following reaction:

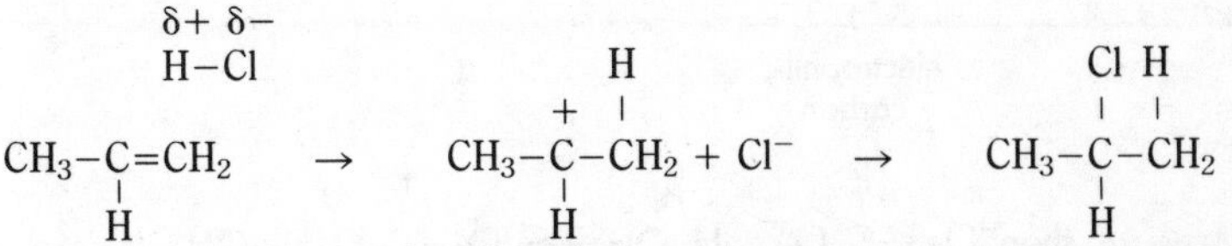

$$
\begin{array}{ccc}
\overset{\delta+\ \delta-}{H-Cl} & H & Cl\ H \\
& +\ | & |\ \ | \\
CH_3-C=CH_2 \ \rightarrow & CH_3-C-CH_2 + Cl^- \ \rightarrow & CH_3-C-CH_2 \\
|\ \ \ \ \ \ \ \ \ \ \ \ \ & | & | \\
H & H & H \\
\text{propene} & \text{a secondary carbocation} & \text{2-chloropropane}
\end{array}
$$

The product is 2-chloropropane (not 1-chloropropane) because the secondary **carbocation** intermediate is more stable than primary and the reaction follows **Markovnikov's rule** – *When a molecule HZ adds to an unsymmetrical alkene, the hydrogen adds to the double bonded carbon with the greater number of H atoms.*

The rapid decolorisation of aqueous bromine (used as a test for alkenes) is considered a fast addition reaction involving bromic(I) acid because BrOH molecules are present in the aqueous bromine equilibrium mixture and 2-bromoethanol ($Br-CH_2-CH_2-OH$) is one of the products:

$$
Br_2(aq) + H_2O(l) \leftrightarrow Br^-(aq) + H^+(aq) + \overset{\delta+\ \delta-}{Br-OH}(aq)
$$

A completely dry gaseous mixture of bromine and ethene does not react in a glass flask whose inner surface is coated with paraffin wax. This suggests that water molecules or ions in the glass surface produce **electrophiles** by polarising some bromine molecules. For instance:

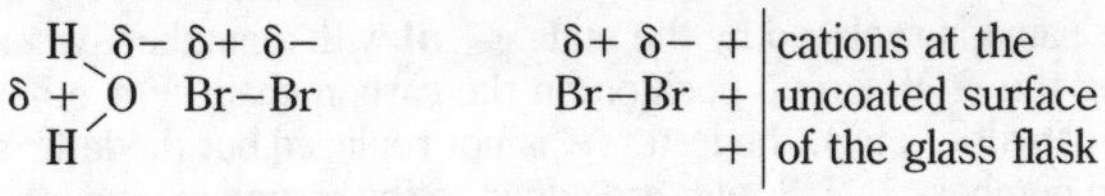

$$
\begin{array}{l|l}
\begin{array}{l}
H\ \ \ \ \delta-\ \ \delta+\ \delta- \\
\delta + \diagdown O\ \ \ \ Br-Br \\
H\diagup
\end{array}
&
\begin{array}{l}
\overset{\delta+\ \ \delta-\ \ +}{Br-Br}\ + \\
\ \ \ \ \ \ \ \ \ \ \ \ \ + \\
\ \ \ \ \ \ \ \ \ \ \ \ \ +
\end{array}
\begin{array}{l}
\text{cations at the} \\
\text{uncoated surface} \\
\text{of the glass flask}
\end{array}
\end{array}
$$

In the presence of lithium chloride (Li^+Cl^-), gaseous ethene reacts with bromine to form some CH_2Cl-CH_2Br as well as CH_2Br-CH_2Br.

Nucleophilic addition reactions are typified by the reactions of **aldehydes** and **ketones** with aqueous sodium hydrogensulphite and with hydrogen cyanide; see Fig. A.16 overleaf. Nucleophilic addition reactions involving the $>C=O$ and $-NH_2$ groups are frequently followed by an **elimination reaction** to give an overall **condensation reaction**.

ADSORPTION

Adsorption is the attachment of molecules to a surface. *Physical adsorption* is a term which usually refers to a layer of gas molecules attached to a solid surface by weak **van der Waals forces**. If the molecules are chemically bonded to the surface, the adsorption is termed *chemisorption*.

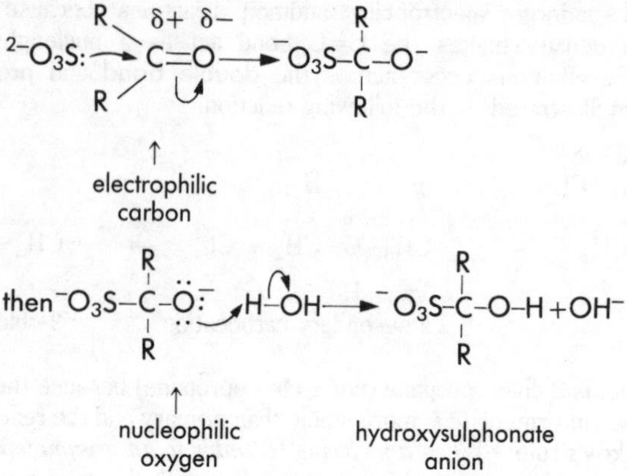

$$2^-O_3S: \overset{R}{\underset{R}{\rightarrow C}} \overset{\delta+ \quad \delta-}{= O} \longrightarrow {}^-O_3S - \overset{R}{\underset{R}{C}} - O^-$$

↑
electrophilic
carbon

$$then\ {}^-O_3S - \overset{R}{\underset{R}{C}} - \overset{..}{\underset{..}{O}}{}^{:-} \overset{\frown}{\ } H - OH \longrightarrow {}^-O_3S - \overset{R}{\underset{R}{C}} - O - H + OH^-$$

↑
nucleophilic
oxygen

hydroxysulphonate
anion

Fig. A.16 Addition reaction with aqueous hydrogensulphite

ALCOHOL

Alcohols are a broad class of organic compounds containing one or more hydroxyl (−OH) groups not attached directly to a benzene ring.

▶ NOMENCLATURE

The root name is based on the longest carbon chain. For alcohols with one −OH group in their molecular structure (monohydric alcohols), the letter 'e' at the end of the root name is replaced by the ending −**ol** with a number, when necessary, showing the −OH group position on the carbon chain. For other alcohols (dihydric, trihydric, etc.), the letter 'e' is not replaced but the letters 'diol', triol, etc. and numbers 1, 2, 3, etc. are added to the ending to show the number and positions of the −OH groups.

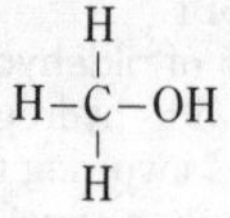

methanol
(methyl alcohol)

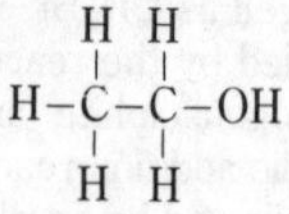

ethanol
(ethyl alcohol)

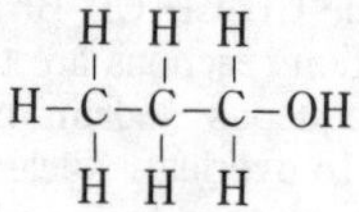

propan-1-ol
(propyl alcohol)

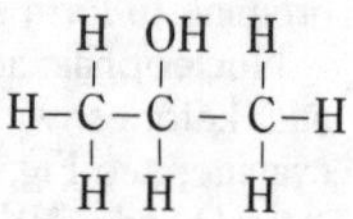

propan-2-ol
(isopropyl alcohol)

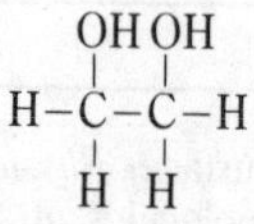

ethane-1,2-diol
(glycol)

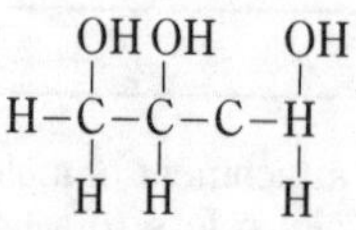

propane-1,2,3-triol
(glycerol or glycerine)

The terms **primary, secondary** and **tertiary** describe the type of alcohol by the alkyl group to which the $-OH$ group is attached.

$$C_3H_7-\underset{\underset{\displaystyle H}{|}}{\overset{\overset{\displaystyle H}{|}}{C}}-OH \qquad C_2H_5-\underset{\underset{\displaystyle CH_3}{|}}{\overset{\overset{\displaystyle H}{|}}{C}}-OH \qquad CH_3-\underset{\underset{\displaystyle CH_3}{|}}{\overset{\overset{\displaystyle CH_3}{|}}{C}}-OH$$

primary   secondary   tertiary

butan-1-ol   butan-2-ol   2-methylpropan-2-ol
(butyl alcohol)   (sec-butyl alcohol)   (tert-butyl alcohol)

Although unsystematic traditional names such as methyl, ethyl, propyl and isopropyl alcohol, glycol and glycerol are found in some textbooks and are still used by some chemists, these are best avoided.

PHYSICAL PROPERTIES

The physical properties are typical of substances with covalent molecular structures, but their melting and boiling points are higher than those of hydrocarbons of similar molar mass, owing to **hydrogen-bonding** and permanent **dipole–dipole forces** between the molecules. The interaction of the $-OH$ groups with water molecules is responsible for the solubility of methanol and ethanol, which mix with water in all proportions, and for the surfactant properties of hexadecan-1-ol (cetyl alcohol), which when spread on lakes lowers the surface tension of water and conserves water by reducing evaporation.

CHEMICAL PROPERTIES

Alcohols combust well in excess air or oxygen to form carbon dioxide and water, unlike phenols which ignite less readily and burn with very smoky flames. Primary alcohols are oxidised by hot aqueous sodium dichromate(VI) acidified with sulphuric acid. If the product is continuously distilled from the reaction mixture at the same rate as the alcohol is dripped onto the oxidising agent, the product is mainly the **aldehyde**:

$$R-\underset{\underset{\displaystyle H}{|}}{\overset{\overset{\displaystyle H}{|}}{C}}-OH \xrightarrow[\substack{\text{continuous}\\\text{distillation}}]{Cr_2O_7^{2-}/H^+} R-\overset{\displaystyle O}{\underset{\displaystyle H}{\overset{\displaystyle \diagup\!\diagup}{C}}\diagdown}$$

primary alcohol   aldehyde

If a mixture of primary alcohol and oxidising agent is refluxed, the product is the **carboxylic acid**:

$$\overset{H}{\underset{H}{R-C-OH}} \quad \xrightarrow[\text{reflux}]{\underset{\text{heat under}}{Cr_2O_7{}^{2-}/14H^+}} \quad R-C\overset{O}{\underset{O-H}{\Big\backslash\!\!\!\diagup}}$$

primary alcohol   carboxylic acid

Secondary alcohols are oxidised to **ketones**:

$$\overset{H}{\underset{R}{R-C-OH}} \quad \xrightarrow[\text{reflux}]{\underset{\text{heat under}}{Cr_2O_7{}^{2-}/14H^+}} \quad R-C\overset{O}{\underset{R}{\Big\backslash\!\!\!\diagup}}$$

secondary alcohol   ketone

Tertiary alcohols oxidise only under drastic conditions.

The $-OH$ group in alcohols reacts with metallic sodium which liberates hydrogen. For example, ethanol is reduced to sodium ethoxide, which dissolves in the excess unreacted ethanol:

$$C_2H_5-O-H(l) + Na(s) \rightarrow C_2H_5-O^-Na^+(alc) + \tfrac{1}{2}H_2(g)$$

This reaction provides a safe way of disposing of sodium residues. The ethoxide ion $(C_2H_5O^-)$ is such a powerful base that its extremely weak conjugate acid, the ethanol molecule, is never considered to be acidic. Aqueous wide-range **indicator** turns purple (**pH 14**) when added to sodium ethoxide:

$$C_2H_5O^-(alc) + H_2O(l) \rightarrow C_2H_5OH(aq) + OH^-(aq)$$

The $-OH$ group in **phenol** reacts similarly but the sodium and the phenol must be molten. **Acyl chlorides** react with the $-OH$ group in alcohols and phenols to produce **esters**:

$$R-O-H + CH_3\overset{O}{\overset{\|}{C}}-Cl \longrightarrow CH_3\overset{O}{\overset{\|}{C}}-O-R + HCl$$

The reactions with **acid anhydrides** are similar but usually less vigorous:

$$R-OH + (CH_3CO)_2O \rightarrow CH_3CO_2R + CH_3CO_2H$$

Esterification of alcohols can be done by direct reaction with carboxylic acids and HCl or H_2SO_4 to supply H^+ as **catalyst**.

The $-OH$ group in alcohols can be replaced by **halogen** by reacting alcohols with reagents such as PCl_5, PCl_3, $SOCl_2$, KBr and H_2SO_4, red phosphorus and iodine. Alcohols containing the structure $CH_3CH(OH)-$ undergo the iodoform reaction by forming a yellow precipitate of CHI_3 on warming with I_2 and NaOH(aq).

Dehydration of alcohols to alkenes and/or ethers can be done by heating with concentrated phosphoric or sulphuric acid or with a **heterogeneous catalyst** such as Al_2O_3. On the industrial scale, alcohols are manufactured by **hydration** of alkenes.

ALDEHYDES

Aldehydes are a class of compound containing the carbonyl **functional group** $>C=O$, with one hydrogen atom always attached to the carbon atom to give the aldehyde functional group $-CHO$.

NOMENCLATURE

The names of the aldehydes end with the suffix **–al** if they are **aliphatic** and **–aldehyde** if they are **aromatic**: for instance, propanal CH_3CH_2CHO and benzenecarbaldehyde C_6H_5CHO. Aldehydes are isomeric with ketones.

PHYSICAL PROPERTIES

The $>C=O$ group gives the aldehydes distinctive **infra-red absorption spectra** and makes the molecules typically polar covalent with permanent **dipole–dipole attractions** contributing to their **intermolecular forces**. Consequently, their melting and boiling points are higher than those of aliphatic **hydrocarbons** and **ethers** but lower than those of aliphatic **alcohols** of similar molar mass; only methanal and ethanal are gases at 25°C and 1 atm. They have the unpleasant smell of rotting fruit. Aldehyde molecules do NOT hydrogen-bond to each other but they will form **hydrogen bonds** with water, alcohol and **carboxylic acid** molecules. So the lower aldehydes are soluble in water but the solubility of the higher aldehydes decreases as the proportion of hydrocarbon in the molecular structure increases. Aqueous methanal (formalin) is a powerful disinfectant that has been used for preserving biological specimens.

CHEMICAL PROPERTIES

Aldehydes burn or explode with excess air or oxygen to carbon dioxide and water. They are readily oxidised to carboxylic acids by warming with acidified aqueous sodium dichromate(VI). For instance:

$$3CH_3-\underset{\underset{H}{|}}{C}=O + Cr_2O_7{}^{2-}(aq) + 8H^+(aq) \rightarrow CH_3-\underset{\underset{OH}{|}}{C}=O + 2Cr^{3+}(aq) + 4H_2O(l)$$

orange solution   green solution

Most aldehydes reduce **Fehling's solution** to an orange-red precipitate of copper(I) oxide on warming. However, on the one hand, C_6H_5CHO

(benzaldehyde or benzenecarbaldehyde) does not reduce Fehling's solution whereas, on the other hand, HCHO (formaldehyde or methanal) reduces the copper(II) complex in the blue solution to metallic copper.

The alkyl groups in an aldehyde molecule can undergo free-radical substitution reactions with chlorine and bromine in ultraviolet light. For example:

$$CH_3-CHO \xrightarrow{\text{Cl}_2 \text{ and UV light}} CCl_3-CHO \quad \text{trichloroethanal (chloral)}$$

A hydrogen atom on the α-carbon atom (the carbon atom next to the C=O group) of an aldehyde is readily replaced by a halogen atom. Ethanal gives a yellow precipitate of CHI_3 ('iodoform') because its molecular structure includes the $CH_3-C=O$ group.

When aldehydes are reduced they form primary alcohols, in contrast to ketones which form secondary alcohols. Reduction by sodium tetrahydridoborate or lithium tetrahydridoaluminate may be seen as i) nucleophilic addition of hydride anions, $:H^-$, supplied by BH_4^- and AlH_4^- ions, to the electrophilic carbon atom in the $>C=O$ group followed by ii) electrophilic attack of hydrogen cations, H^+, supplied by the water molecules, upon the nucleophilic oxygen atom in the carbonyl group (see Fig. A.17).

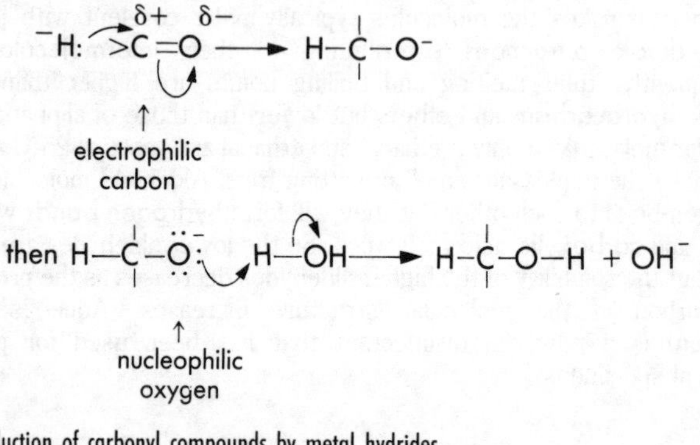

Fig.A.17 Reduction of carbonyl compounds by metal hydrides

Aldehydes undergo nucleophilic addition reactions with HCN as NaCN(aq) followed by HCl(aq) and with $NaHSO_3$(aq) or Na_2SO_3(aq). They also undergo *addition—elimination* reactions with ammonia and its derivatives, those with ammonia often being complicated and giving resinous condensation polymers. For example, methanal reacts with ammonia to form water and hexamethylenetetramine, a heterocyclic fused ring compound used industrially in corrosion inhibitors, fungicides, rubber-to-textile adhesives, shrink-proofing textiles and high-explosive cyclonite:

$$6HCHO + 4NH_3 \rightarrow 6H_2O + (CH_2)_6N_4$$

These reactions are explained by the polarity of the C=O bond making the C-atom an electrophilic centre and by the O-atom being able to accept a displaced π-electron pair to form a fairly stable intermediate oxoanion. These reactions are typical of carbonyl compounds: see Figs. C.11 and C.13. In contrast to ketones, aldehydes readily polymerise to form resins.

ALICYCLIC COMPOUND

Alicyclic compounds are organic compounds in which the carbon atoms form rings but not the **aromatic** rings of benzene and its derivatives. **Cycloalkanes** and cycloalkenes are two important classes of **alicyclic** compound, having the same general formula as **alkenes** and **alkynes** respectively.

ALIPHATIC COMPOUND

Aliphatic compounds are organic compounds in which the carbon atoms form straight or branched chains but do not form rings. **Alkanes, alkenes** and **alkynes** are the three major classes of aliphatic or **acyclic hydrocarbons.**

ALKALI

An alkali is a base that dissolves in water to give a solution with a greater concentration of hydroxide ion than hydrogen ion. The **hydroxides** and **oxides** of the **s-block metals** are strong alkalis. Ammonia and the lower **aliphatic amines** are weak alkalis.

ALKALI METALS

The alkali metals form a set of closely similar elements (Group I) and part of the **s-block** in the **periodic table**; see Fig. A.18. The properties of the elements are governed by the one electron in the outer s-**orbital** of their atoms. Each element has a characteristic line **emission spectrum** and a distinctive flame colour.

▶ PHYSICAL PROPERTIES

The alkali metals are ductile, malleable, and good conductors of heat and electricity. Compared to the **alkaline earth metals** of Group II and to most **transition metals,** they are soft and have low densities, melting points, boiling points and standard enthalpy changes of melting and of boiling, all of which decrease down the group with increasing atomic number. All the alkali metals have a **body-centred cubic structure** and their **metallic bonding** is weaker than that in most transition metals.

▶ CHEMICAL PROPERTIES

The alkali metals are powerful **reducing agents,** losing the outer s-subshell

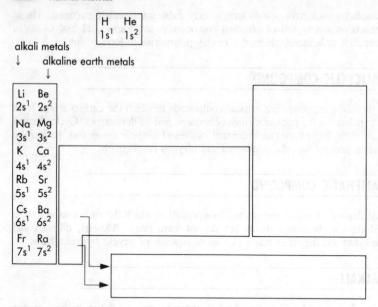

Fig. A.18 s-block elements

electron to form very stable cations in an oxidation state of +1 with the same electronic configurations as their corresponding **noble gases**; for example:

$$Na(2,8,1) \rightarrow Na^+(2,8) + e^- \qquad [Ne(2,8)]$$

They are more reactive than their corresponding alkaline earth metals, burning vigorously in air to form normal **oxides** (Li_2O), peroxides (Na_2O_2) or superoxides (KO_2), depending upon their reactivity. The metals react violently with halogens and with sulphur to form ionic binary halides whose aqueous solutions are good electrolytic conductors – for example:

$$2Na(s) + Cl_2(g) \xrightarrow{\text{heat}} 2NaCl(s)$$

In the UK, sodium is used as the reducing agent in the industrial extraction of titanium from $TiCl_4$:

$$4Na + TiCl_4 \xrightarrow{1000°C} 4NaCl + Ti$$

Alkali metals combine, on heating, with hydrogen to form ionic hydrides that react readily with water:

$$2Li(s) + H_2(g) \xrightarrow{\text{heat}} 2LiH(s)$$

$$LiH(s) + H_2O(l) \longrightarrow LiOH(aq) + H_2(g)$$

The alkali metals react with water to form hydrogen and strongly basic hydroxides, the reaction becoming more violent with increasing atomic number down the group:

$$2Na(s) + 2H_2O(l) \rightarrow 2Na^+(aq) + 2OH^-(aq) + H_2(g)$$

The alkali metal cations are so stable that they are not normally discharged at the **cathode** during **electrolysis** of their aqueous solutions. Consequently, the metals must be extracted by electrolysis of their molten anhydrous salts. For example, metallic sodium is manufactured by electrolysis of molten sodium chloride (to which calcium chloride and/or barium is added to lower the melting point). The applied voltage is high enough to discharge sodium ions but not the more stable barium and calcium ions.

$$Na^+ + e^- \xrightarrow[\text{molten } Na^+Cl^-]{\text{at the cathode}} Na$$

As the first member of Group I, lithium shows some a typical properties and a **diagonal relationship** with magnesium in Group II. For example, lithium combines directly with nitrogen to form lithium nitride:

$$6Li(s) + N_2(g) \rightarrow 2Li_3N(s)$$

Lithium hydroxide and lithium carbonate decompose into lithium oxide when heated in a bunsen flame:

$$Li_2CO_3(s) \rightarrow Li_2O(s) + CO_2(g)$$

ALKALINE

Alkaline is a term which refers to **aqueous** solutions in which the **concentration** of the **hydroxide ion**, $OH^-(aq)$, is greater than the concentration of the **hydrogen ion**, $H_3O^+(aq)$ and at 25°C the **pH** is greater than 7.

ALKALINE EARTH METAL

The alkaline earth metals form a set of reactive elements (Group II) and part of the **s-block** in the **periodic table**: see Fig. A.18. The properties of the elements are governed by the two electrons in the outer s-**orbital** of their atoms. Calcium, strontium and barium have distinctive flame colours.

▶ PHYSICAL PROPERTIES

The alkaline earth metals are ductile, malleable, and good conductors of heat and electricity. They have higher densities, melting points, boiling points and standard enthalpy changes of melting and of boiling than the **alkali metals** but, compared to most **transition metals**, the alkaline earth metals are soft and have low densities. Their **metallic bonding** is stronger than that of the alkali metals but weaker than the bonding in most transition metals. Hardness, density, standard enthalpy change of melting and of boiling all decrease down the group with increasing atomic number. The alkaline earth metal structures, melting points and boiling points vary irregularly in the group: see Fig. A.19.

Symbol	Structure	m.pt/°C	b.pt/°C
Li	BCC	181	1342
Na	BCC	98	883
K	BCC	63	760
Rb	BCC	39	686
Cs	BCC	29	669
Be	HCP	1278	2970
Mg	HCP	649	1107
Ca	FCC	839	1484
Sr	FCC	769	1384
Ba	BCC	725	1640

BCC body-centred cubic; HCP hexagonal close packed; FCC face-centred cubic

Fig. A.19 Physical properties of alkaline earth metals

▶ CHEMICAL PROPERTIES

These metals act as powerful **reducing agents** by losing their outer s-subshell electrons to form very stable cations in oxidation state +2 with the same **electronic configurations** as their corresponding **noble gases**. For instance:

$$Mg(2,8,2) \rightarrow Mg^{2+}(2,8)+2e^- \quad [Ne\ 2,8]$$

Alkaline earth metals are less reactive than their corresponding alkali metals. They burn in air to form **oxides**, peroxides and nitrides, depending upon their reactivity:

$$Ba(s) + O_2(g) \rightarrow BaO_2(s) \quad [barium\ peroxide]$$

They react, often violently, with **halogens** and with sulphur to form halides and sulphides:

$$Mg(s) + S(s) \xrightarrow{\ heat\ } MgS(s)$$

Outside the UK, magnesium is used as the reducing agent in the industrial extraction of titanium from $TiCl_4$:

$$2Mg + TiCl_4 \xrightarrow{\ 1000°C\ } 2MgCl_2 + Ti$$

The metals combine, on heating, with hydrogen to form ionic **hydrides** that react readily with water:

$$Ca(s) + H_2(g) \xrightarrow{\text{heat}} CaH_2(s)$$

$$CaH_2(s) + 2H_2O(l) \longrightarrow Ca(OH)_2(s) + 2H_2(g)$$

The reaction of the metals with water becomes more vigorous with increasing atomic number down the group. Magnesium must be heated in steam but barium reacts readily with cold water:

$$Mg(s) + H_2O(g) \xrightarrow{\text{heat}} MgO(s) + H_2(g)$$

The metals are extracted by **electrolysis** of their molten anhydrous salts because their aqueous cations are too stable to discharge at the **cathode** during electrolysis of their aqueous solutions; the less stable aqueous hydrogen ion discharges instead. Magnesium is the most important Group II metal on account of its strength, low density, high heat resistance and its ability to form alloys. Calcium carbonate and oxide are the most important Group II compounds because of their widespread use in agriculture, building materials and the production of iron.

ALKANE

The alkanes are a **homologous series** of **saturated hydrocarbons** with the general formula $C_nH_{(2n+2)}$. These compounds used to be called the paraffins (Latin *para* – little; *affinis* – reaction). They are a major constituent of petroleum, from which they are obtained by **distillation** and **fractional distillation**.

▶ NOMENCLATURE

The names of the alkanes are very important because they form the basis of the names of many organic compounds. The first four **straight-chain** alkanes are known by their traditional names. The other members of the homologous series of alkanes have names based on Greek or latin words for the numbers; see Fig. A.20.

Number of carbons	Molecular formula	Name	Number of carbons	Molecular formula	Name
1	CH_4	methane	11	$C_{11}H_{24}$	undecane
2	C_2H_6	ethane	12	$C_{12}H_{26}$	dodecane
3	C_3H_8	propane	13	$C_{13}H_{28}$	tridecane
4	C_4H_{10}	butane	14	$C_{14}H_{30}$	tetradecane
5	C_5H_{12}	pentane	15	$C_{15}H_{32}$	pentadecane
6	C_6H_{14}	hexane	18	$C_{18}H_{38}$	octadecane
7	C_7H_{16}	heptane	20	$C_{20}H_{42}$	eicosane
8	C_8H_{18}	octane	21	$C_{21}H_{44}$	heneicosane
9	C_9H_{20}	nonane	30	$C_{30}H_{62}$	triacontane
10	$C_{10}H_{22}$	decane	100	$C_{100}H_{202}$	hectane

Fig. A.20 Names of some straight-chain alkanes

The branched-chain isomers are named according to rules laid down by IUPAC (International Union of Pure and Applied Chemistry); see Fig. A.21.

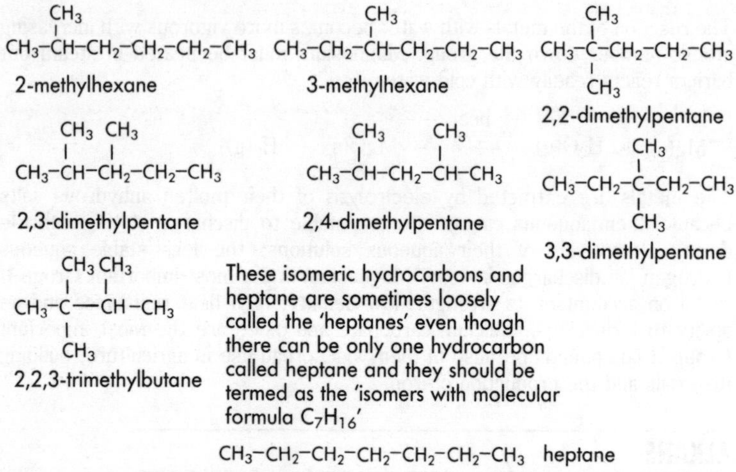

$$CH_3-\underset{\underset{CH_3}{|}}{CH}-CH_2-CH_2-CH_2-CH_3$$

2-methylhexane

$$CH_3-CH_2-\underset{\underset{CH_3}{|}}{CH}-CH_2-CH_2-CH_3$$

3-methylhexane

$$CH_3-\underset{\underset{CH_3}{|}}{\overset{\overset{CH_3}{|}}{C}}-CH_2-CH_2-CH_3$$

2,2-dimethylpentane

$$CH_3-\underset{\underset{CH_3}{|}}{CH}-\underset{\underset{CH_3}{|}}{CH}-CH_2-CH_3$$

2,3-dimethylpentane

$$CH_3-\underset{\underset{CH_3}{|}}{CH}-CH_2-\underset{\underset{CH_3}{|}}{CH}-CH_3$$

2,4-dimethylpentane

$$CH_3-CH_2-\underset{\underset{CH_3}{|}}{\overset{\overset{CH_3}{|}}{C}}-CH_2-CH_3$$

3,3-dimethylpentane

$$CH_3-\underset{\underset{CH_3}{|}}{\overset{\overset{CH_3}{|}}{C}}-\underset{\overset{|}{CH_3}}{\overset{CH_3}{|}}{CH}-CH_3$$

2,2,3-trimethylbutane

These isomeric hydrocarbons and heptane are sometimes loosely called the 'heptanes' even though there can be only one hydrocarbon called heptane and they should be termed as the 'isomers with molecular formula C_7H_{16}'

$$CH_3-CH_2-CH_2-CH_2-CH_2-CH_2-CH_3 \quad \text{heptane}$$

Fig. A.21 Naming isomers of heptane

The root name is based on the longest continous carbon chain and the name of the straight-chain alkane with the same number of carbons. The root is prefixed with names based on the shorter carbon branches and the names of the corresponding straight-chain alkanes. The number of identical branches (two, three, four, etc.) is indicated by the appropriate adjunct (**di–**, **tri–**, **tetra–**, etc.). The positions of the branches on the longest chain are numbered from the end giving the lower number for the initial branching point. The prefixes are attached in alphabetical order of branch name and ignoring adjuncts such as **di–** and **tri–**. The naming procedure is illustrated in Fig. A.22.

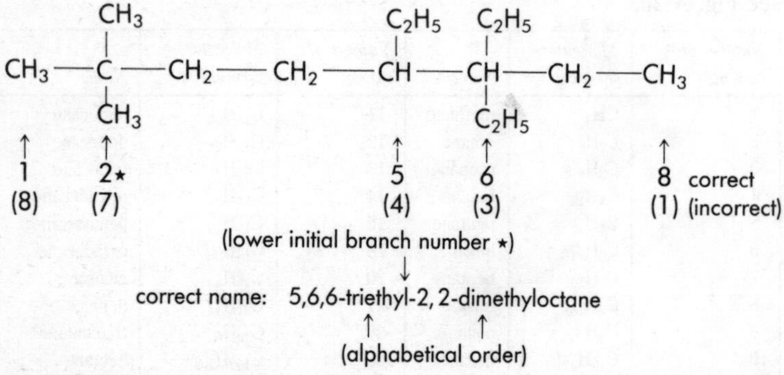

$$CH_3-\underset{\underset{CH_3}{|}}{\overset{\overset{CH_3}{|}}{C}}-CH_2-CH_2-\underset{\underset{C_2H_5}{|}}{\overset{\overset{C_2H_5}{|}}{CH}}-\underset{}{\overset{\overset{C_2H_5}{|}}{CH}}-CH_2-CH_3$$

| 1 | 2* | | 5 | 6 | | 8 correct |
| (8) | (7) | | (4) | (3) | | (1) (incorrect) |

(lower initial branch number *)

correct name: 5,6,6-triethyl-2,2-dimethyloctane

(alphabetical order)

Fig. A.22 Naming a hydrocarbon

A new homologous series of organic compounds is obtained simply by changing the ending (suffix) –**ane**: e.g. alkanes→**alkenes**→**alkynes**→ alkanols→alkanals→alkanones→alkanoic acids. Great care must be taken with the spelling of the names of organic compounds.

▶ REACTIONS OF ALKANES

As fuels, alkanes combust to water and carbon dioxide or carbon monoxide if there is insufficient air:

$$C_nH_{(2n+2)} + \tfrac{1}{2}(3n+1)O_2 \rightarrow nCO_2 + (n+1)H_2O$$

They react spontaneously and explosively with fluorine and will burn in chlorine. Alkanes also undergo substitution reactions with chlorine and bromine. The gas phase reaction of chlorine with methane is an example of an organic **substitution reaction** involving a **free-radical chain mechanism**. When a red-brown solution of bromine in hexane is exposed to light from the sun or a photoflood lamp, steamy fumes of hydrogen bromide are given off and the red-brown colour fades. A variety of bromohexanes may be identified in the liquid product by **gas chromatography**.

$$C_6H_{14}(l) + Br_2(l) \xrightarrow{\text{light}} C_6H_{13}Br(l) + HBr(g)$$

Large alkane molecules are broken down into smaller ones and into alkenes by thermal and catalytic **cracking** on the industrial scale.

ALKENE

The alkenes are a **homologous series** of **acyclic unsaturated hydrocarbons** with the general formula C_nH_{2n}. These compounds used to be called the *olefins* ('the oil formers') because their **addition reactions** with **halogens** give oily products. On the industrial scale, alkenes are produced by **cracking** and **dehydrogenation** of ethane, propane and other **alkanes** from natural gas and light **petroleum** distillates in the naphtha range; for instance:

$$CH_3-CH_3 \xrightarrow{-H_2} CH_2=CH_2$$

In the laboratory, impure samples of alkenes can be made by the **dehydration** of **alcohols** or the dehydrobromination of bromoalkanes – for instance:

$$CH_3-CH_2OH - H_2O \xrightarrow[\text{OR pass vapour over hot Al}_2O_3 \text{ or SiO}_2]{\text{heat liquid with excess H}_2SO_4 \text{ or H}_3PO_4} CH_2=CH_2$$

$$CH_3-CH_2Br - HBr \xrightarrow[\text{OR pass vapour over hot soda-lime}]{\text{heat liquid KOH in ethanolic solution}} CH_2=CH_2$$

▶ **NOMENCLATURE**

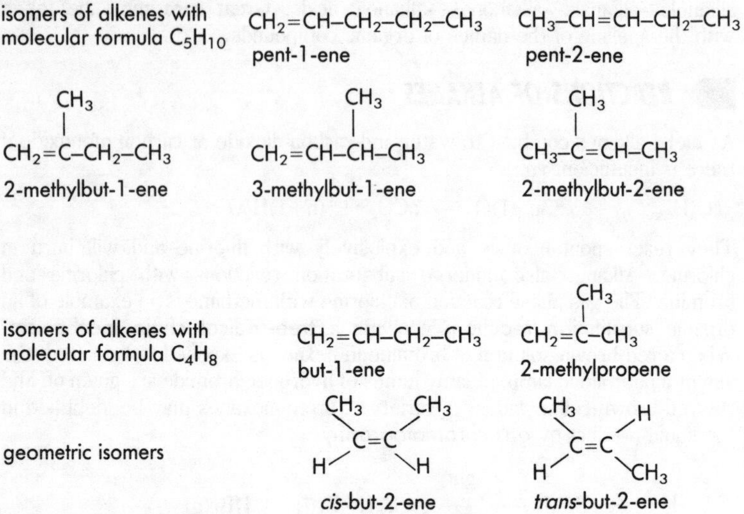

isomers of alkenes with
molecular formula C_5H_{10}

$CH_2=CH–CH_2–CH_2–CH_3$
pent-1-ene

$CH_3–CH=CH–CH_2–CH_3$
pent-2-ene

$$CH_2=\overset{\overset{\displaystyle CH_3}{\displaystyle |}}{C}–CH_2–CH_3$$
2-methylbut-1-ene

$$CH_2=CH–\overset{\overset{\displaystyle CH_3}{\displaystyle |}}{CH}–CH_3$$
3-methylbut-1-ene

$$CH_3–\overset{\overset{\displaystyle CH_3}{\displaystyle |}}{C}=CH–CH_3$$
2-methylbut-2-ene

isomers of alkenes with
molecular formula C_4H_8

$CH_2=CH–CH_2–CH_3$
but-1-ene

$$CH_2=\overset{\overset{\displaystyle CH_3}{\displaystyle |}}{C}–CH_3$$
2-methylpropene

geometric isomers

$$\underset{H}{\overset{CH_3}{}}C=C\underset{H}{\overset{CH_3}{}}$$
cis-but-2-ene

$$\underset{H}{\overset{CH_3}{}}C=C\underset{CH_3}{\overset{H}{}}$$
trans-but-2-ene

Fig. A.23 Naming isomeric alkenes

The root name of an alkene is based on the name of the **straight-chain alkane** with the same number of carbon atoms as the longest carbon chain containing the **double bond**. The prefixes **cis–** and **trans–** are used to name simple **geometric isomers** having the same group on each of two doubly bonded carbon atoms; see Fig. A.23. Alkenes have nasty smells and burn with smoky flames. They act as **nucleophiles** and undergo addition reactions with **electrophiles**. They rapidly decolorise red-brown aqueous bromine and turn purple aqueous potassium manganate(VII) to a clear, colourless solution when acidic and to a cloudy, brown suspension when alkaline:

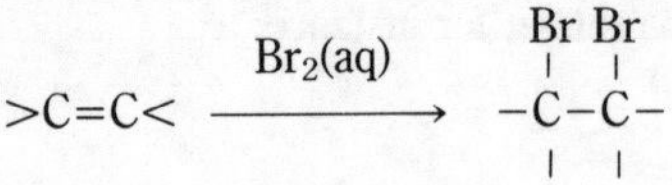

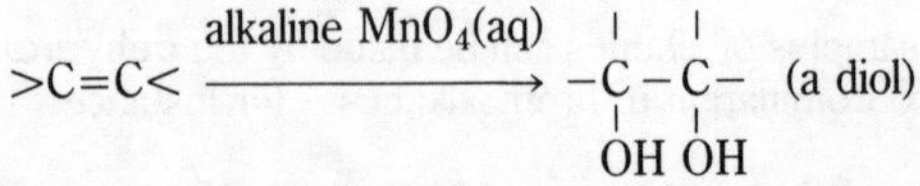

$$>C=C< \xrightarrow{\;Br_2(aq)\;} \overset{\overset{\displaystyle Br\;\;Br}{\displaystyle |\;\;\;|}}{\underset{\displaystyle |\;\;\;|}{-C-C-}}$$

$$>C=C< \xrightarrow{\;alkaline\;MnO_4(aq)\;} \underset{\overset{\displaystyle |\;\;\;\;|}{OH\;\;OH}}{\overset{\displaystyle |\;\;\;\;|}{-C-C-}}\;\;(a\;diol)$$

The rapid decolorisation of aqueous bromine is used as a test for alkenes. Alkenes can be **hydrated**, by reaction with sulphuric acid and then water, to form alcohols – for instance:

$$CH_3CH=CH_2(g) + H_2O(l) \xrightarrow{\;concentrated\;H_2SO_4\;} CH_3CH(OH)CH_3$$

The addition occurs across the double bond according to **Markovnikov's** rule to give propan-2-ol and not propan-1-ol. Propene $CH_3-CH=CH_2$ is a typical alkene whose electrophilic addition reactions are summarised in Fig. A.15.

Hydrogen and a nickel catalyst can be used to hydrogenate alkenes. The addition reaction reduces the **double bond** in the unsaturated alkenes to a **single bond** in the corresponding **saturated hydrocarbons**: see Fig. A.24.

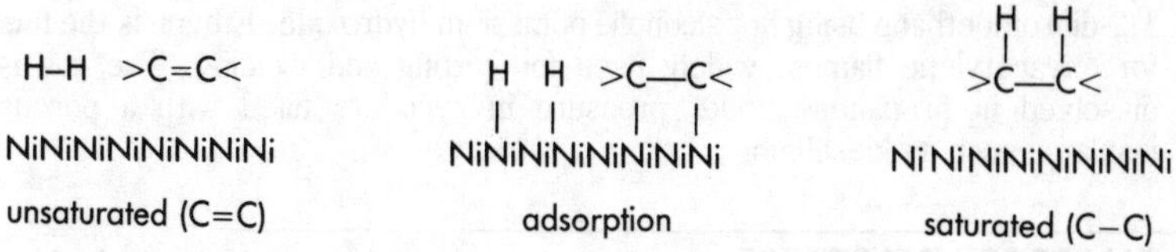

$$H-H \quad >C=C<$$
NiNiNiNiNiNiNiNi
unsaturated (C=C)

$$H \quad H \quad >C-C<$$
NiNiNiNiNiNiNiNi
adsorption

$$\begin{array}{c} H \quad H \\ | \quad | \\ >C-C< \end{array}$$
Ni NiNiNiNiNiNiNi
saturated (C-C)

Unsaturated edible oils are hydrogenated to saturated edible fats in the production of margarine: e.g.

$$CH_3(CH_2)_7CH=CH(CH_2)_7CO_2CH_2$$
$$CH_3(CH_2)_7CH=CH(CH_2)_7CO_2CH \xrightarrow[\text{Ni catalyst}]{3H_2(g)} $$
$$CH_3(CH_2)_7CH=CH(CH_2)_7CO_2CH_2$$

$$CH_3(CH_2)_7CH_2-CH_2(CH_2)_7CO_2CH_2$$
$$CH_3(CH_2)_7CH_2-CH_2(CH_2)_7CO_2CH$$
$$CH_3(CH_2)_7CH_2-CH_2(CH_2)_7CO_2CH_2$$

olein (from olive oil) an ester of
propane-1,2,3-triol and
octadec-9-enoic (oleic) acid

stearin (in most fats) an ester of
propane-1,2,3-triol and
octadecanoic (stearic) acid

Fig. A.24 Hydrogenation of alkenes

Alkenes will polymerise at high-pressure by a **free-radical chain mechanism** to form branched chain stereo-irregular (atactic) addition **polymers** with low crystallinity, density and rigidity or at low-pressure by a mechanism involving **Ziegler–Natta catalysts** to form stereo-regular (isotactic and syndiotactic) addition polymers with high crystallinity, density and rigidity:

$$n >C=C< \longrightarrow (-\overset{|}{\underset{|}{C}}-\overset{|}{\underset{|}{C}}-)_n$$

Addition polymerisation of alkenes is an extremely important industrial process.

ALKYLATION

Alkylation is the introduction of an alkyl group into a molecule, usually by substituting a hydrogen atom. Alkylation of the benzene ring in **aromatic** compounds may be performed by a **Friedel–Craft reaction** using an alkyl chloride and anhydrous aluminium chloride as a **catalyst**:

$$C_6H_6 + CH_3Cl \xrightarrow[\text{catalyst}]{AlCl_3} C_6H_5CH_3 + HCl$$

ALKYNE

Alkynes are a **homologous series** of **unsaturated acyclic hydrocarbons** having the general formula $C_nH_{(2n-2)}$. The simplest alkyne is ethyne (acetylene): $H-C\equiv C-H$. It is a colourless gas which can be prepared by the **hydrolysis** of calcium dicarbide or by the **dehydrohalogenation** of 1,2-dibromoethane using hot alcoholic potassium hydroxide. Ethyne is the fuel for oxyacetylene flames, widely used for cutting and welding. The gas is dissolved in propanone under pressure in cylinders filled with a porous insulator such as kieselguhr.

ALLOTROPY/ALLOTROPE

Allotropy is the existence of two or more different forms of the same element in the same physical state. Allotropes are normally different crystalline forms of a **p-block element** from Groups IV, V and VI in the **periodic table**. They have different physical and/or chemical properties; see Fig. A.25.

Diamond and graphite are well known allotropes of carbon. *Enantiotropy* is a form of allotropy exhibited by sulphur, for example, in which each allotrope is stable over a definite temperature range and both allotropes (rhombic and monoclinic) can coexist in stable **equilibrium** at a definite temperature (96°C) known as the **transition temperature**. Oxygen (O_2) and ozone (O_3) are gaseous allotropes.

Red phosphorus	*White phosphorus*
violet-red amorphous powder	transparent yellowish brittle waxy solid
density $2.34\,g\,cm^{-3}$	density $1.82\,g\,cm^{-3}$
sublimes at 416 °C	melts under water at 44.1 °C
giant covalent molecules	simple covalent molecules (P_4)
insoluble in organic solvents	soluble in organic solvents
does not react with NaOH(aq)	reacts with NaOH(aq) $\rightarrow$ $PH_3(g)$ + $NaH_2PO_4(aq)$

Fig. A.25 Two allotropes of phosphorus

ALPHA AMINO ACID

◀ Amino acid ▶

ALPHA PARTICLE

An α-particle is helium **nucleus**, consisting of two **protons** and two **neutrons**, emitted by an **atom** during radioactive decay. When an atom undergoes alpha decay, its **mass number** decreases by 4 and its **atomic number** decreases by 2 as, for example, when uranium-235 decays to thorium-231:

$$^{235}_{92}U \rightarrow {}^{231}_{90}Th + \alpha^{2+}$$

A value for the **Avogadro constant** is obtained by i) detecting and counting in a measured solid angle the α-particles emitted by a radioactive source, ii) collecting all the α-particles emitted by the source over a period of time, iii) measuring the volume of helium collected and iv) calculating the number of helium atoms there would be in a **mole** of helium.

AMIDE

◀ Acid amides ▶

AMINE

Amines are classes of organic **base** formally derived from ammonia.

▶ NOMENCLATURE

Amines are classed as primary, secondary or tertiary according to the number of hydrogen atoms in an ammonia molecule that have been replaced by alkyl or aryl groups:

CH_3NH_2

methylamine
(primary)

$(CH_3)_2NH$

dimethylamine
(secondary)

$(CH_3)_3N$

trimethylamine
(tertiary)

The 'simple' amines above are named according to the groups attached to the N-atom. The names of 'more complicated' primary amines are based on the name of the hydrocarbon and $-NH_2$ as the amino group:

$CH_3CH_2CH_2NH_2$

1-aminopropane

$CH_3CHNH_2CH_3$

2-aminopropane

▶ PHYSICAL PROPERTIES

Methylamine and ethylamine smell like rotting fish. Butylamine (l-aminobutane) $C_4H_9NH_2H$ and phenylamine (aniline) $C_6H_5NH_2$ with their polar covalent molecular structures show the typical physical properties of **aliphatic** and **aromatic** primary amines. The electronegative N-atom and its **lone** electron **pair** are responsible for the molecular polarity and trigonal pyramidal shape which in turn result in permanent **dipole–dipole** intermolecular attractions. Primary and secondary amines are also **hydrogen-bonded**. These dipole–dipole and hydrogen bonding forces account for the solubility of amines in water, alcohol and ethoxyethane and for their higher boiling points compared to **hydrocarbons** of comparable molar mass.

▶ CHEMICAL PROPERTIES

Many amines are hazardous. They are toxic when absorbed through the skin and highly flammable, burning in excess air or oxygen to form carbon dioxide, water and nitrogen – for instance:

$$2C_3H_7NH_2(l) + 10\tfrac{1}{2}O_2(g) \rightarrow 6CO_2(g) + 9H_2O(l) + 2N_2(g)$$

Most amines are **weak bases** in aqueous solution, with primary alkylamines like butylamine ($pK_b = 3.2$) being stronger than ammonia ($pK_b = 4.8$) because the electron-donating alkyl group C_4H_9 makes the butylammonium cation more stable than the ammonium cation:

$$\begin{array}{c} H \\ | \\ C_4H_9-N: \\ | \\ H \end{array} + H_2O \longrightarrow \begin{array}{c} H \\ | \\ C_4H_9-N^+-: \\ | \\ H \end{array} + OH^-$$

Phenylamine ($pK_b = 9.4$) is a weaker base than ammonia because the lone pair of electrons on the N-atom participate in the delocalized π-electron system of the benzene ring, stabilising the free base molecule and making the lone electron pair less readily available for dative bonding. So the phenylammonium cation is less readily formed; see Fig. A.26.

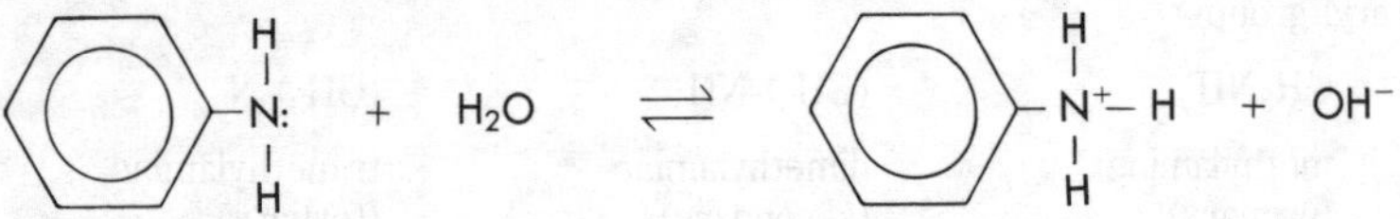

Fig. A.26 Phenylamine is a weaker base than ammonia

Amines, like ammonia, turn moist red litmus paper blue and form a white smoke with hydrogen chloride:

$$R-NH_2(g) + HCl(g) \rightarrow R-NH_3^+Cl^-(s) \text{ white smoke of a substituted}$$
ammonium chloride

They dissolve in aqueous acids to form solutions of their substituted ammonium salts from which they can crystallise as white solids. The fishy smell disappears as the amine dissolves:

$$C_6H_5NH_2(l) + HCl(aq) \rightarrow C_6H_5NH_3^+(aq) + Cl^-(aq)$$

The smell returns if a strong base is added to neutralise the acid and displace the weaker amine base:

$$C_6H_5NH_3^+(aq) + OH^-(aq) \rightarrow C_6H_5NH_2(l) + H_2O(l)$$

Amines, like ammonia, act as ligands to **d-block cations** to form **complex ions** whose solutions are often blue or green, for instance:

$$Cu^{2+}(aq) + 4 :N-R(aq) \rightarrow [Cu(:N-R)_4]^{2+}(aq)$$

Phenylamine forms an insoluble green complex with aqueous copper(II) sulphate. **EDTA** (ethylenediaminetetraacetic acid) is an important complexing agent involving two tertiary amino groups. The nitrogen atom in the amine **functional group** acts as a **Lewis base** by donating its lone electron pair to form a **dative bond** with the central ion.

The lone pair also makes the N-atom a nucleophilic centre so that amines react with **halogenoalkanes, acyl chlorides** and **acid anhydrides**. This explains why the reaction of ammonia with, for example, bromoethane forms a mixture of primary, secondary and tertiary ethylamines and/or the quaternary ethylammonium bromide salt and is not usually used to synthesise amines:

$$NH_3 + C_2H_5Br \rightarrow C_2H_5NH_2 + HBr \text{ OR } C_2H_5NH_3^+Br^-$$

$$C_2H_5NH_2 + C_2H_5Br \rightarrow (C_2H_5)_2NH + HBr \text{ OR } (C_2H_5)_2NH_2^+Br^-$$

$$(C_2H_5)_2NH + C_2H_5Br \rightarrow (C_2H_5)_3N + HBr \text{ OR } (C_2H_5)_3NH^+Br^-$$

$$(C_2H_5)_3N + C_2H_5Br \rightarrow (C_2H_5)_4N^+ + Br^- \text{ OR } (C_2H_5)_4N^+Br^-$$

The reaction of primary and secondary amines with acyl chlorides and acid anhydrides forms substituted amides but does not usually produce a mixture because the N-atom is far less nucleophilic in the amide than in the original amine (see Fig. A.27). This acylation of an amine functional group is often used to protect it from reagents attacking the rest of the molecule as, for example, when preparing nitrophenylamine (See Fig. A.13). Primary aliphatic amines react with nitrous acid to give highly unstable aqueous diazonium salts which decompose to give various products and, almost quantitatively, one mole of N_2 gas for every mole of $-NH_2$ group. The **diazotisation** reaction of aromatic primary amines with ice-cold nitrous acid is less complicated and gives stable aqueous diazonium salts that are important intermediates in the synthesis of **azo-dyes** and other aromatic compounds.

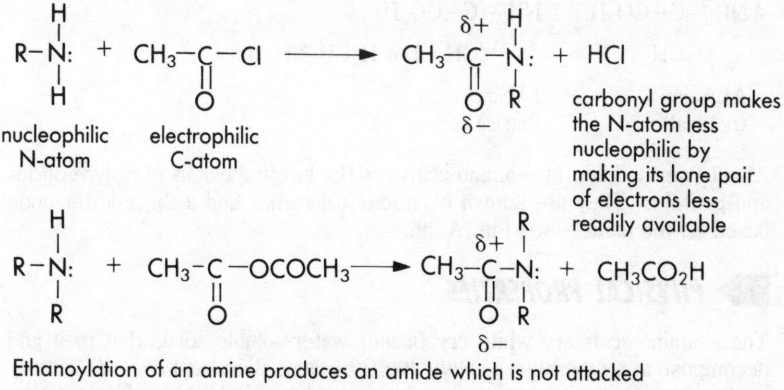

carbonyl group makes the N-atom less nucleophilic by making its lone pair of electrons less readily available

Ethanoylation of an amine produces an amide which is not attacked

Fig. A.27 Amides formed by acylation of amines

AMMINE

Ammines are **complexes** in which the **ligands** are ammonia molecules. If excess concentrated aqueous ammonia is added to aqueous copper(II) sulphate, the initially formed blue precipitate of copper hydroxide dissolves to form a deep blue solution containing the octahedral tetraamminediaquacopper(II) complex cation $[Cu(NH_3)_4(H_2O)_2]^{2+}$.

AMINO ACID

An α-amino acid is an organic compound containing an amino group ($-NH_2$), a carboxyl group ($-CO_2H$), a hydrogen atom ($-H$) and a 'side-chain' (R) consisting of an atom or group (such as $-H$, $-CH_3$, etc.), all attached to the same carbon atom:

$$NH_2-\overset{\overset{\displaystyle H}{|}}{\underset{\underset{\displaystyle R}{|}}{C}}-CO_2H$$

The amino acids are classified as acidic, neutral or basic according to the relative number of $-CO_2H$ and $-NH_2$ groups in the molecule.

$$NH_2-\overset{\overset{\displaystyle H}{|}}{\underset{\underset{\displaystyle CH_2CH_2CO_2H}{|}}{C}}-CO_2H \qquad NH_2-\overset{\overset{\displaystyle H}{|}}{\underset{\underset{\displaystyle H}{|}}{C}}-CO_2H$$

GLUtamic acid   GLYcine
(acidic)   (neutral)

$$NH_2-\overset{\overset{\displaystyle H}{|}}{\underset{\underset{\displaystyle CH_3}{|}}{C}}-CO_2H \qquad NH_2-\overset{\overset{\displaystyle H}{|}}{\underset{\underset{\displaystyle CH_2CH_2CH_2CH_2NH_2}{|}}{C}}-CO_2H$$

ALAnine   LYSine
(neutral)   (basic)

About twenty different α-amino-acids are the building blocks of **polypeptides** and **proteins**. They are known by traditional names and a three-letter code based on the name – see Fig. A.29.

▶ PHYSICAL PROPERTIES

The α-amino-acids are white crystalline, water-soluble solids that melt and decompose at temperatures above 200°C because they exist as inner salts or **zwitterions** (German *zwitter* – hybrid) formed by **proton transfer** from the $-CO_2H$ group to the $-NH_2$ group:

$$\overset{\overset{\displaystyle NH_2}{|}}{\underset{\underset{\displaystyle H}{|}}{R-C}}-CO_2H \quad \rightarrow \quad \overset{\overset{\displaystyle NH_3^+}{|}}{\underset{\underset{\displaystyle H}{|}}{R-C}}-CO_2^-$$

GLY	glycine	$-H$		ASP	aspartic acid	$-CH_2CO_2H$
ALA	alanine	$-CH_3$		GLU	glutamic acid	$-CH_2CH_2CO_2H$
VAL	valine	$-CH(CH_3)_2$		SER	serine	$-CH_2OH$
LEU	leucine	$-CH_2CH(CH_3)_2$		THR	threonine	$-CH(OH)CH_3$
ILE	isoleucine	$-CH(CH_3)CH_2CH_3$		TYR	tyrosine	$-CH_2C_6H_4OH$
PHE	phenylalanine	$-CH_2C_6H_5$		ASN	asparginine	$-CH_2CONH_2$
LYS	lysine	$-CH_2CH_2CH_2CH_2NH_2$		GLN	glutamine	$-CH_2CH_2CONH_2$
CYS	cysteine	$-CH_2SH$		MET	methionine	$-CH_2CH_2SCH_3$
ARG	arginine	$-CH_2CH_2CH_2NHC=NH$		HIS	histidine	$-CH_2-C=CH$

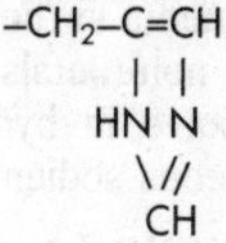

TRP tryptophan

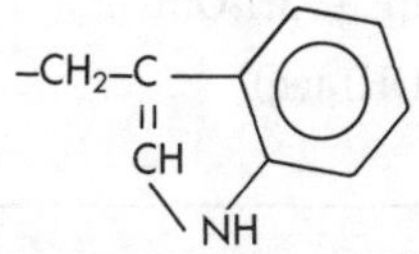

PRO proline

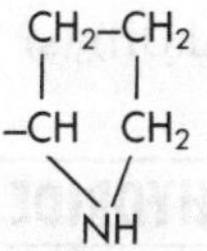

Fig. A.28 α-amino acids are characterised by their side-chain

Except in glycine, the assymetric α-carbon is a chiral centre, so α-amino-acids can exist as **enantiomers** with identical physical properties, except for their effect upon the **plane of polarised light**, and with identical chemical properties, except for their reaction with another chiral molecule (see Fig. A.29). It is actually the L-form of the α-amino-acids that is the building-block of all proteins. Amino acids in a mixture, obtained by the **hydrolysis** of protein, can be separated by **paper chromatography**, revealed by spraying and heating with ninhydrin and identified by their R_f values.

L (+) glutamic acid

D (−) glutamic acid

letter ↑ specifies the form of the structure

sign ↑ specifies the direction of rotation of the plane of polarised light

Fig. A.29 Enantiomers of glutamic acid

AMOUNT OF SUBSTANCE

The mole is the amount of substance containing as many atoms, molecules, ions or other entities as specified by a **formula**, as there are carbon atoms in 12g of the carbon **isotope**, ^{12}C.

AMPHOTERIC OXIDE

An amphoteric oxide is an oxide that can react as an acid with bases and as a base with acids. The formation of amphoteric oxides is often regarded as a characteristic of metalloids. Amphoteric oxides are formed by elements such as boron, aluminium, silicon, germanium, arsenic and antimony: that is, elements in the periodic table close to the diagonal dividing the metals from the non-metals. For example, freshly prepared aluminium hydroxide will dissolve in hydrochloric acid to form aqueous aluminium chloride and in aqueous sodium hydroxide to form aqueous sodium tetrahydroxyaluminate:

$$Al(OH)_3(s) + 3HCl(aq) \rightarrow AlCl_3(aq) + 3H_2O(l)$$

$$Al(OH)_3(s) + NaOH(aq) \rightarrow NaAl(OH)_4(aq)$$

ANHYDRIDE

An anhydride is a compound formally derived from an acid by the removal of a water molecule. Sulphur trioxide (SO_3) is the anhydride of sulphuric acid (H_2SO_4).
◀ Acid anhydride ▶

ANION

An anion is a negatively charged atom or group of atoms. The charge on a simple anion may be related to the electronic configuration of the atom and the position of the element in the periodic table. Simple anions carrying a charge of 3^-, 2^- and 1^- are formed by non-metallic elements in Groups V, VI and VII respectively. During electrolysis, anions such as the halide ions, the oxide ion (O^{2-}) and the hydroxide ion (OH^-) move towards the positive electrode (anode) where they may be discharged to form the non-metallic element.

ANISOTROPIC

An anisotropic substance is a crystalline solid whose refractive index varies with the direction of the light through its crystal. Most substances are anisotropic because their crystal structures are different when viewed from different directions. In ionic compounds such as calcite, $CaCO_3$, these differences arise because the crystal structure does not belong to the cubic crystal system and it contains non-spherical ions: the carbonate anion is trigonal planar. Other properties of anisotropic crystals may also depend upon the direction in which they are measured: for example, the thermal conductivity of pyrographite is much greater along the crystal planes than at right-angles to them.

ANODE

The anode is the positively charged **electrode** in an **electrolytic** cell. In many industrial electrolytic manufacturing processes, such as the production of chlorine by the electrolysis of brine, the anode is made of graphite. In industrial electrolytic refining of metals such as copper, the anode consists of the impure metal.

AQUEOUS

Aqueous means that a substance is dissolved in water. The aqueous state is shown in formulae and equations by the letters **aq**. Thus NaCl(aq) means a solution of sodium chloride in water.

ARENE

Arenes are **hydrocarbons** containing one or more benzene rings. These hydrocarbons and their derivatives are also known as aromatic hydrocarbons because they were first prepared from compounds which had fragrant aromas and which were extracted from plant oils. Benzene (C_6H_6) is the simplest arene. The **functional group** (C_6H_5-) derived from it is called the phenyl group. The names of the alkylbenzene hydrocarbons include the names of the alkyl groups and numbers (from 1 to 6) indicating where the groups are attached to the benzene ring; see Fig. A.30.

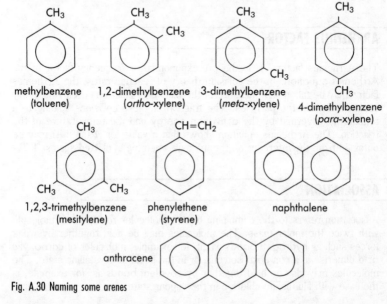

methylbenzene
(toluene)

1,2-dimethylbenzene
(*ortho*-xylene)

3-dimethylbenzene
(*meta*-xylene)

4-dimethylbenzene
(*para*-xylene)

1,2,3-trimethylbenzene
(mesitylene)

phenylethene
(styrene)

naphthalene

anthracene

Fig. A.30 Naming some arenes

Arenes undergo predominantly **electrophilic substitution** reactions even though the arenes are more **unsaturated** than the **alkenes** which usually undergo **electrophilic addition** reactions. The substitution reactions allow the benzene ring to preserve its extra stability caused by the **delocalisation** of the π-electrons.

AROMATIC

The term aromatic refers to a major class of unsaturated cyclic organic compounds which includes benzene, naphthalene, anthracene, phenanthrene and their derivatives.
◀ Arenes ▶

ARRHENIUS EQUATION

The Arrhenius equation $k = A_e^{-(E_a/RT)}$ is a mathematical expression for the dependence of the **rate constant** (k) upon the absolute temperature.

The Arrhenius equation may be written in the form of a logarithmic relationship $\ln k = \ln A - E_a/RT$ where R is the **gas constant**, A is the Arrhenius (or pre-exponential) factor and E_a is the **activation energy** for the reaction. This form of the equation shows that the activation energy for a reaction may be determined from the gradient $(= -E_a/R)$ of a graph of $\ln k$ against $1/T$. Note that $e^{-(E_a/RT)}$ may be seen as a measure of the fraction of molecular collisions resulting in a reaction.

ARRHENIUS FACTOR

The Arrhenius factor, A, is the pre-exponential or frequency factor in the **Arrhenius equation**. In the collision theory of reaction rates, the Arrhenius factor may be taken as a measure of the billions of collisions occurring every second between the particles. The fraction of these collisions leading to a reaction is governed by the **activation energy** and the temperature of the reaction. The Arrhenius equation shows that a value for A (as $\ln A$) may be obtained from the intercept on the $\ln k$ axis when $\ln k$ is plotted against $1/T$.

ASSOCIATION

Association refers to the combining of two molecules to form one molecule with twice the molar mass. The molecules may be held together by weak forces such as **hydrogen bonds** when, for example, molecules of **carboxylic acid** dimerise in a non-polar solvent, or in the liquid or crystalline state. The molecules may also be held together by **covalent bonds** as, for example, is the case with aluminium chloride in the vapour state:

$$\begin{array}{c} O\cdots H-O \\ CH_3C \diagup \qquad \diagdown CCH_3 \\ O-H\cdots O \end{array}$$

$$\begin{array}{ccc} Cl & Cl & Cl \\ \diagdown \diagup \diagdown \diagup \\ Al & Al \\ \diagup \diagdown \diagup \diagdown \\ Cl & Cl & Cl \end{array}$$

ASYMMETRIC CARBON

An asymmetric carbon **atom** is a carbon atom to which four different atoms or **groups** of atoms are attached. Apart from glycine, the $\propto$-carbon atom in the naturally occurring **amino acids** is asymmetric and acts as the chiral centre responsible for the **optical activity** of the molecule. Molecules containing an asymmetric carbon atom can exist as **enantiomers**.

ATACTIC

An atactic **polymer** is usually formed by **free radical polymerisation** as a sticky material with very low crystallinity resulting from a random arrangment of side groups (X) along the polymer's main carbon chain.

$$-\overset{X}{\underset{}{C}}-\overset{X}{\underset{X}{C}}-\overset{X}{\underset{}{C}}-C-C-\overset{X}{\underset{X}{C}}-C-C-\overset{X}{\underset{X}{C}}-C-C-C-C-C-C-\overset{X}{\underset{}{C}}-C-C-\overset{X}{\underset{X}{C}}-C-C-\overset{X}{\underset{}{C}}-C-C-\overset{X}{\underset{}{C}}-C-$$

The irregular structures of atactic polymers make them less useful than the **isotactic** and **syndiotactic** polymers with their regular structures.

ATOM

An atom is the smallest particle of an element, consisting of an extremely small **nucleus** of **protons** and **neutrons** surrounded by a large volume of space which, apart from the **electrons**, is empty. Diagrams of atoms are rarely drawn to scale because the nucleus is too small compared to the space occupied by the electrons. The nucleus constitutes most of the mass of the atom. All atoms of the same element must have the same **atomic number** (Z) but may exist as **isotopes** with different **mass numbers** (A) – see Fig. A.31. The nuclei of radioactive elements disintegrate spontaneously to emit either

α-particles or β-particles and sometimes γ-radiation.

The electrons surrounding the nucleus are arranged on the basis of their energies in **orbitals**. The **electronic configuration** represents, by a sequence of numbers, letters and superscripts, the grouping of orbitals into shells and subshells.

At one time scientists believed that the submicroscopic particles of elements could not be split (Greek *atmos* – indivisible). In 1909 **Geiger and Marsden** experimented on the scattering of α-particles, and in 1911 Ernest Rutherford proposed his theory of the nuclear atom to account for their

carbon isotope $^{13}_{6}C$: atomic number 6 and mass number 13

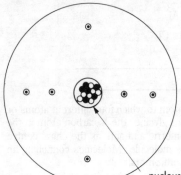

name	relative mass	relative charge
⊙ electron	1	−1
● proton	1836	+1
○ neutron	1836	0

nucleus (not to scale): 6 protons and 7 neutrons

Fig. A.31 An isotope of carbon

results. In 1913 **Moseley's** work on **X-rays** related the atomic number to the number of protons in the nucleus, and in 1932 Chadwick discovered the neutron. Recent research in theoretical and high-energy physics has shown that atoms are composed of many different subatomic particles.

ATOMIC EMISSION SPECTRUM

energy to move electron completely out of H-atom from ground state level is the ionisation energy = 1312 kJ mol^{-1}

Balmer series:
frequency: 0.46 to 0.79 × 10^{15} Hz
lines in the visible
region of the spectrum

Lyman series:
frequency: 2.47 to 3.29 × 10^{15} Hz
lines in the 'invisible'
ultra violet region of the spectrum

increasing levels of electron energy

∞
5
4
3
2
1
ground state level

1312 kJ

Fig. A.32 Levels of electron energy in a hydrogen atom

An atomic emission spectrum is a line spectrum produced when the atoms or ions of an element are excited by electrical or thermal energy and the light emitted is observed through a spectroscope. Each line in the spectrum corresponds to electromagnetic radiation of a particular frequency and therefore of a definite energy emitted by electrons in the atom changing from higher to lower levels of energy. The lines in the atomic emission spectrum of hydrogen form several series and can be related to differences in electron energy levels within the atom – see Fig. A.32.

Values for levels of electron energy can be obtained from atomic spectra. The convergence limit of the **Lyman series** can be used to calculate the **ionisation energy** of hydrogen.

ATOMIC NUMBER

Atomic number (Z) is the number of **protons** in the **nucleus** of an atom. It is also the numbered position of the element in the **periodic table**.

ATOMIC RADIUS

An atomic radius is a measure of the size of an **atom**. The atomic radius is not a fixed rigid value for each element. The value depends upon the nature and number of the surrounding atoms. The metallic radius of an atom is one-half of the distance between the centres of two adjacent atoms in a close-packed **crystal structure**. The **covalent radius** is one-half of the distance between the nuclei of two atoms covalently bonded in the same molecule. The *van der Waals radius* is one-half of the distance between the nuclei of two atoms in adjacent molecules – see Fig. A.33.

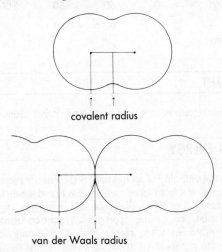

covalent radius

van der Waals radius

Fig. A.33 Atomic radii

Atomic radii increase with increasing atomic number down a group as the increasing number of complete inner shells of electrons repel the outer electrons and shield them from the attraction of the increasing nuclear charge. Atomic radii decrease with increasing atomic number across a period in the periodic table because the **inner shell shielding** remains fairly constant so the increasing nuclear charge pulls the outer electrons closer to the nucleus.

ATOMIC VOLUME

Atomic volume is the mass of one **mole** of atoms of an element divided by the **density** of the element in the liquid or solid state. In 1870 Julius Lothar Meyer published his plot of atomic volume against atomic weight (relative atomic mass) and showed that this physical property obeyed the **periodic law**; see Fig. A.34.

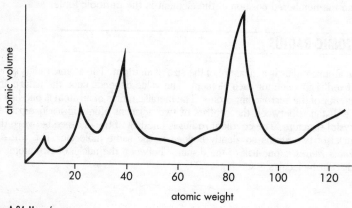

Fig. A.34 Meyer's curves

ATOMIC WEIGHT

Atomic weight is the almost obsolete term for **relative atomic mass**.

ATOMISATION ENERGY

The standard molar **enthalpy** of atomisation is the energy required to produce a mole of gaseous atoms from an element at 298 K and a constant pressure of 1 atm. The atomisation energies of carbon and hydrogen, together with the enthalpies of combustion of the appropriate **hydrocarbons**, are used to calculate bond **enthalpies** using Hess's law. The atomisation energies of sodium and chlorine are included in the data required to calculate the **lattice energy** of sodium chloride using the Born–Haber cycle.

AUTOCATALYSIS

An autocatalytic reaction is a reaction in which one of the products acts as a catalyst for the reaction itself.

The iodination of propanone is autocatalytic because the reaction is first-order with respect to hydrogen ions and hydrogen ions are formed during the reaction:

$$CH_3COCH_3(aq) + I_2(aq) \rightarrow CH_3COCH_2I(aq) + H^+(aq) + I^-(aq)$$

The oxidation of ethanedioic acid by aqueous manganate(VII) ions and sulphuric acid is autocatalysed by the manganese(II) ions formed by the reduction of the manganate(VII) ions:

$$5(CO_2H)_2(aq) + 2MnO_4^-(aq) + 6H^+(aq) \rightarrow$$
$$2Mn^{2+}(aq) + 8H_2O(l) + 10CO_2(g)$$

In an autocatalytic reaction, the rate of change of concentration of a reactant will start low, increase to a maximum and then decrease to zero. This contrasts with other reactions in which the rate of the reaction will start high and decrease to zero.

AVOGADRO CONSTANT

The Avogadro constant (L) is the proportionality constant connecting the number of entities (specified by a **formula**) to the amount of substance (expressed in moles):

number of entities = $L \times$ amount of substance

The Avogadro constant has units (mol^{-1}) and should not be confused with the Avogadro number, the name given to the (pure) number of atoms in 12 g of carbon-12. Values for the constant have been determined by various methods including experiments on radioactive decay, electrolysis and X-ray diffraction. The currently accepted value of the Avogadro constant is, to four significant figures, 6.022×10^{23} mol^{-1},

AVOGADRO'S PRINCIPLE

Equal volumes of gases at the same temperature and pressure contain the same number of molecules.

The Italian chemist Amadeo Avogadro stated his principle in 1811 and provided an explanation for Gay–Lussac's law of combining volumes. One consequence of Avogadro's principle is that the molar volume of any gas is approximately constant. An alternative statement of the principle is that the volume of a gas, at a constant temperature and pressure, is directly proportional to its amount in moles: $V \propto n$.

AZEOTROPES

An azeotrope is a homogeneous mixture of two liquids that boils at a constant temperature, giving a distillate of constant composition. Azeotrope comes from a Greek word meaning 'not changed on boiling'. Azeotropes (or azeotropic mixtures) are often called constant boiling mixtures. They are solutions which deviate so much from **Raoult's law** that their boiling point/composition diagrams show a maximum or a minimum (see Fig. A.35). A solution of 96 per cent ethanol and 4 per cent water forms a minimum boiling azeotrope with a positive deviation from Raoult's law. A solution of 68 per cent nitric acid and 32 per cent water forms a maximum boiling azeotrope with a negative deviation from Raoult's law.

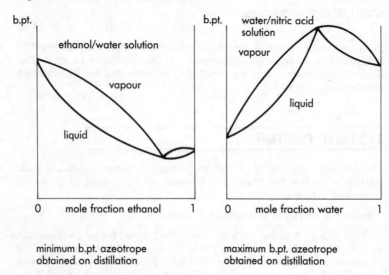

minimum b.pt. azeotrope obtained on distillation

maximum b.pt. azeotrope obtained on distillation

Fig. A.35 Distillation of binary azeotropes

The components of azeotropic mixtures cannot be separated by **distillation**. The boiling point and the composition of an azeotrope varies with pressure.

AZO DYE

An azo dye is a coloured organic compound containing the group $-N=N-$ as part of its molecular structure.

Azo dyes are made by **diazotisation** of **aromatic** primary amino groups followed by reaction of the diazonium ions with suitable coupling agents; see Fig. A.36. Diazotisation followed by **coupling** with naphthalen-2-ol to produce an insoluble red precipitate is a simple diagnostic test for aromatic primary amines.

1 Make an ice-cold solution of aniline in hydrochloric acid:

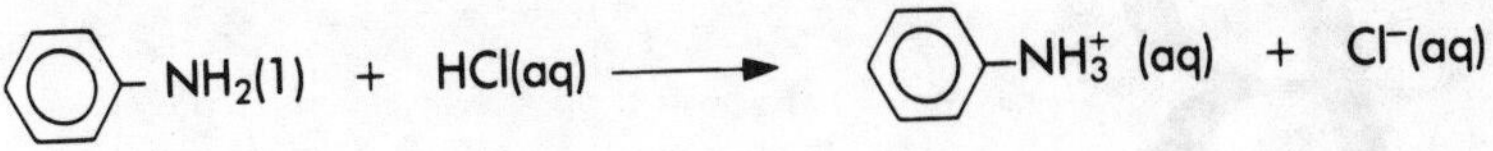

2 Make an ice-cold solution of sodium nitrite in water:

$$NaNO_2(s) \rightarrow Na^+(aq) + NO_2^-(aq)$$

3 Add the aqueous sodium nitrite to the phenylammonium chloride solution drop by drop with vigorous stirring and adding pieces of ice to the mixture as required to keep the solution cold:

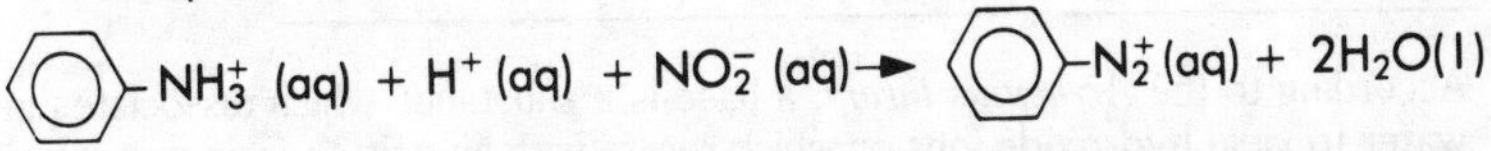

This step to form the aqueous benzene dazonium ion is called the diazotisation reaction.

4 Make an ice-cold solution of napthalen-2-ol (2-naphthol or α-naphthol) in aqueous sodium hydroxide:

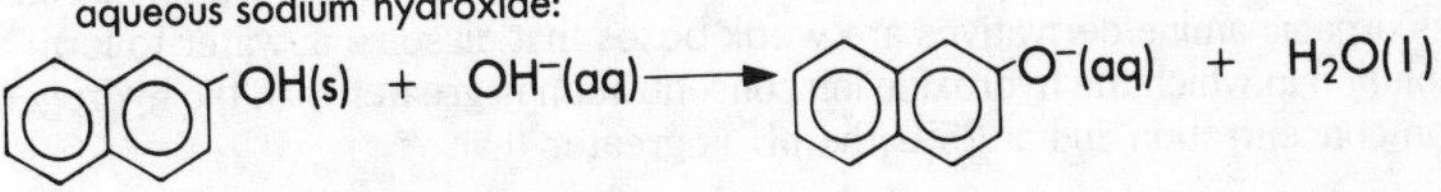

This solution is called the coupling agent.

5 Mix together the ice-cold solutions of benzene diazonium ion and coupling agent to produce a scarlet precipitate of the azo dye by an electrophilic substitution called a coupling reaction:

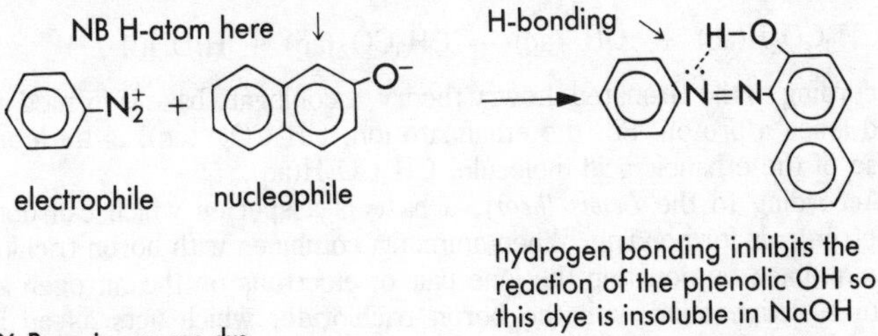

NB H-atom here ↓

H-bonding

electrophile   nucleophile

hydrogen bonding inhibits the reaction of the phenolic OH so this dye is insoluble in NaOH

Fig. A.36 Preparing an azo dye

◀ Qualitative analysis ▶

BASE

According to the *Arrhenius theory*, a base is a substance which dissociates in water to yield **hydroxide** ions or which reacts with an **acid** to form water and a salt.

The ionic hydroxides of the Group I metals and some of the Group II metals are water-soluble strong bases known as **alkalis**. The **oxides** and hydroxides of many other metals are insoluble bases that react with acids. Ammonia and its organic amine derivatives are **weak bases** that dissolve in water to form a solution in which the hydroxide ion concentration is greater than the hydrogen ion concentration and at 25°C the pH is greater than 7:

$$NH_3(aq) + H_2O(l) \rightleftharpoons NH_4^+(aq) + OH^-(aq)$$

According to the **Bronsted–Lowry theory**, a base is a species that can accept a **proton**. The hydroxide ion is a base that can accept a proton from an acid by using a **lone electron pair** on the oxygen to form a **dative bond** with the proton:

$$CH_3CO_2H(aq) + :OH^-(aq) \rightarrow CH_3CO_2^-(aq) + H:OH(l)$$

According to the Bronsted–Lowry theory, a conjugate base is formed when an acid loses a proton. So, the ethanoate ion, $CH_3CO_2^-$ (aq), is the **conjugate base** of the ethanoic acid molecule, $CH_3CO_2H(aq)$.

According to the *Lewis theory*, a base is a species which can donate an electron pair for bonding. When ammonia combines with boron trichloride, it acts as base by donating the lone pair of electrons on the nitrogen atom to form a dative bond with the boron trichloride, which acts as an acid by receiving the electron pair:

$$NH_3(g) + BCl_3(g) \rightarrow H_3N:BCl_3(s)$$

BASE DISSOCIATION CONSTANT

The dissociation constant, K_b, of a **base**, BOH, is the **equilibrium constant** for the reaction $BOH(aq) \rightleftharpoons OH^-(aq) + B^+(aq)$ as given by the expression:

$$\frac{[OH^-(aq)][B^+(aq)]}{[BOH(aq)]} = K_b$$

K_b measures the strength of a base on a scale from 1×10^0 mol dm^{-3} (for the OH$^-$(aq) ion) to 1×10^{-14} mol dm^{-3} (for the H$_2$O(l) molecule). Ammonia and organic amines such as ethylamine are typically moderately **weak bases** with K_b values of around 1×10^{-5} mol dm^{-3}. The dissociation constant of a base (K_b) and the dissociation constant of its **conjugate acid** (K_a) are related to the **ionic product** of water (K_w) by the expression:

$$K_a \times K_b = 1 \times 10^{-14} \text{ mol}^2 \text{ dm}^{-6} \text{ (at 25°C)}$$

Consequently, values for K_b may be calculated from tables of values for the acid dissociation constant, K_a; see Fig. A.7. For example, the conjugate acid of ammonia is the ammonium ion, NH$_4^+$. K_a for the ammonium ion is 5.6×10^{-10} mol dm^{-3}. Therefore, K_b for ammonia will be given by $K_w/K_a = 10^{-14}/5.6 \times 10^{-10}$ or 1.8×10^{-5} mol dm^{-3}.

BENZENE

Benzene (C$_6$H$_6$) is the simplest **arene** or aromatic hydrocarbon. Benzene melts at 5.6°C to form a colourless volatile liquid which boils at 80.2°C and 1 atm. It is an extremely dangerous carcinogen which can be absorbed through the skin. Inhalation of even very low concentrations of the vapour can produce chronic effects. It is highly flammable and the vapour forms explosive mixtures with air. Benzene is banned in most schools and practical work is usually carried out with methylbenzene (toluene), which is less toxic.

▶ STRUCTURE

Benzene is a flat molecule. Its six carbon and six hydrogen atoms are all in the same plane. The C—H bond length (0.108 nm) is normal for hydrocarbons but the distance between the centres of neighbouring carbon atoms (0.139 nm) is longer than a C=C **double bond** (0.134 nm) and shorter than a C—C **single bond** (0.154 nm). The benzene molecule does not have three double bonds, so the structure should be represented as in Fig. B.1.

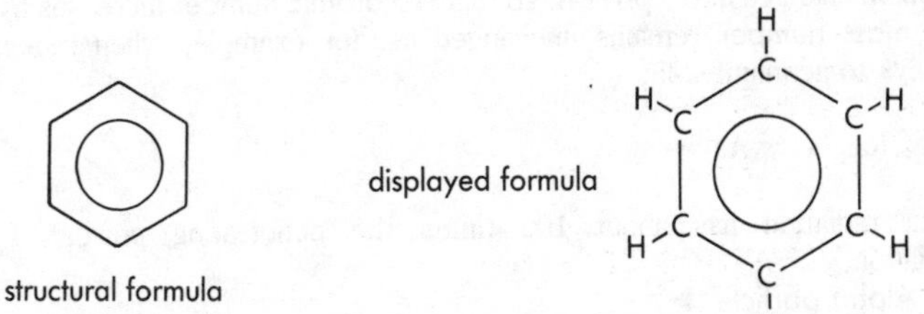

structural formula   displayed formula

Fig. B.1 Delocalised π-electrons in a benzene molecule

The **delocalisation** of π-electrons (represented by the circle in Fig. B.1) confers extra stability on the benzene molecule. Consequently, the ring undergoes substitution rather than **addition reactions** in order that the products keep the extra stability of π-electron delocalisation. The delocalised electron cloud above and below the plane of the ring makes the benzene molecule nucleophilic. Consequently, benzene undergoes predominantly **electrophilic substitution** reactions such as alkylation with **Friedel–Craft reagents**, chlorination and bromination with a **halogen-carrier** catalyst, nitration and sulphonation – see Fig. B.2.

Electrophile	Reagents and conditions	Substitution	Product
Cl^+	Cl_2 and anhydrous $AlCl_3$	 Cl	chlorobenzene
CH_3^+	CH_3Cl and anhydrous $AlCl_3$	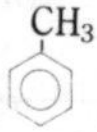 CH_3	methylbenzene
CH_3CO^+	CH_3COCl and anhydrous $AlCl_3$	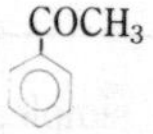 $COCH_3$	phenylethanone
Br^+	Br_2 and Fe or $FeBr_3$	 Br	bromobenzene
NO_2^+	conc. HNO_3 and H_2SO_4 at 45 °C	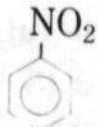 NO_2	nitrobenzene
SO_3	SO_3 in H_2SO_4 (fuming sulphuric) or conc. sulphuric under reflux	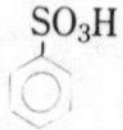 SO_3H	benzenesulphonic acid

Fig. B.2 Electrophilic substitution reactions of benzene

BETA PARTICLE

A β-particle is a high-energy **electron** emitted by an **atom** during radioactive decay. When an atom undergoes beta decay, an electron is emitted and a **neutron** changes into a **proton**, so that the **atomic number** increases by 1 but the **mass number** remains unchanged as, for example, when radium–225 decays to actinium–225:

$$^{225}_{88}\text{Ra} \rightarrow {}^{225}_{89}\text{Ac} + e^-$$

Beta radiation has about 100 times the penetrating power of alpha radiation.
◀ Alpha particle ▶

BIDENTATE LIGAND

A bidentate ligand is a molecule or anion that forms two **dative bonds** with the central atom or cation in a **complex**.

Ethane-1,2-diamine is an example of a bidentate molecule and the (oxalato) ethanedioate ion is an example of a bidentate ion:

$$
\begin{array}{ll}
NH_2 & \quad O \quad\; O- \\
\;\;\backslash CH_2 & \quad\;\; \backslash\;/ \\
\;\;\;\;| & \quad\;\;\; C \\
\;\;/ CH_2 & \quad\;\;\; | \\
NH_2 & \quad\;\;\; C \\
& \quad\;\; /\; \backslash \\
& \quad O \quad\; O-
\end{array}
$$

Lone electron pairs on the nitrogen and oxygen atoms are donated for bonding.

BIMOLECULAR REACTION

A bimolecular reaction is a step in a chemical reaction that involves two species.

If the rate of a reaction which proceeds by a series of steps is governed by a slow, rate-determining bimolecular step, the overall chemical reaction is sometimes called a bimolecular reaction. However, it is better to restrict the use of the terms bimolecular and unimolecular to the *individual steps* in a reaction. For example, the rate of iodination of aqueous propanone is governed by the bimolecular reaction step between one propanone molecule and one hydrogen ion:

$$(CH_3)_2CO(aq) + H_3O^+(aq) \rightarrow (CH_3)_2COH^+(aq) + H_2O(l)$$

The molecularity of a reaction must not be confused with the order of a reaction. The terms bimolecular reaction and second-order reaction have different meanings.

BIOCHEMISTRY

Biochemistry is the study of the chemistry of living organisms. Techniques such as chromatography, X-ray crystallography and electron microscopy have helped to provide information on the structure and function of carbohydrates, lipids, nucleic acids and proteins.

BODY-CENTRED CUBIC STRUCTURE

The body-centred cubic structure is one of the three types of crystal structure, adopted particularly by the alkali metals: see Fig. B.3. The atoms in a body-centred structure have a coordination number of 8. The structure is not close-packed compared to the cubic and hexagonal close-packed structures. This accounts for the softness and low density of the alkali metals compared to most other metals.

◄ Close packing ►

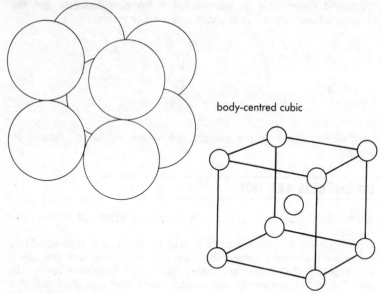

body-centred cubic

Fig. B.3 Structure of alkali metals

BOILING POINT

The boiling point of a liquid is the temperature at which the **vapour pressure** of the liquid equals the pressure of the surrounding atmosphere. Since the temperature at which a liquid boils will vary with the external pressure, the normal boiling point of a liquid is recorded in data books for a standard external pressure of 1 atm (101 325 Pa).

The boiling point is an important physical property; it serves to characterise a substance and is monitored during the purification of a substance by **distillation** or **fractional distillation**.

BOMB CALORIMETER

A bomb calorimeter measures heats of combustion at constant volume. The bomb is a stainless steel vessel which is filled with pure oxygen at high pressure to ensure rapid and complete combustion. The bomb is immersed in a jacket of water that is stirred to maintain an even temperature. The heat capacity of the calorimeter may be found (calibrated) by measuring the temperature rise produced by an accurately known amount of heat. This heat may be supplied either by burning in the bomb a measured amount of a substance (such as benzoic acid) whose heat of combustion is well known or by heating the bomb with an accurately measured amount of electrical energy. See Fig. B.4.

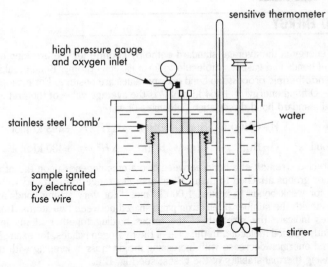

sensitive thermometer

high pressure gauge
and oxygen inlet

stainless steel 'bomb'

water

sample ignited
by electrical
fuse wire

stirrer

Fig. B.4   heat of complete combustion measured at constant volume

◀ Enthalpy changes ▶

BOND ANGLE

A bond angle is the angle between two bonds joining together three **atoms** in a **molecule** or ion. Bond angles can be determined by a technique such as X-ray diffraction. They are taken as the angle between two lines joining the centres of bonded atoms — see Fig. B.5. The values recorded in data books are average values because the angles are influenced by the molecular environment of the three atoms and because they are constantly fluctuating with the motion of the molecules. Bond angles in many simple molecules can be fairly accurately predicted by the **valence shell electron pair repulsion (VSEPR)** theory.

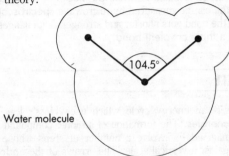

104.5°

Water molecule

Fig. B.5 Bond angle in a water molecule

BOND ENERGY

Bond energy is the average standard **enthalpy change** for the breaking of a mole of bonds in a gaseous **molecule** to form gaseous **atoms**. Bond breaking is an **endothermic process**, so bond energy values are positive. For example, the H−O bond energy of $+464$ kJ mol^{-1} is the average value of the first and second standard bond dissociation enthalpies of water:

first $\quad\quad$ H−OH(g) $\rightarrow$ H(g) + OH(g); $\quad \triangle H^{\theta} = +498$ kJ mol^{-1}

second $\quad\quad$ O−H(g) $\rightarrow$ O(g) + H(g); $\quad \triangle H^{\theta} = +430$ kJ mol^{-1}

The precise strength of a bond between two atoms depends upon the other atoms or groups attached to them. Bond energies range from around 150 kJ mol^{-1} for weak bonds to around 1,000 kJ mol^{-1} for very strong bonds and increase with the number of electron pairs shared between two atoms. Bond energies indicate the strength of the forces holding together atoms in a covalently bonded molecule. In the case of the hydrogen halides, for example, the bond energies decrease with increasing **molar mass** in keeping with the decreasing thermal stability of the gases; see Fig. B.6.

Bond	$E(X-Y)$ /kJ mol^{-1}	Bond	$E(X-Y)$ /kJ mol^{-1}	Bond	$E(X-Y)$ /kJ mol^{-1}	Bond	$E(X-Y)$ /kJ mol^{-1}
HO−OH	144	C−C	347	C=C	612	H−F	568
I−I	151	N−H	391	C≡C	838	H−Cl	432
F−F	158	C−H	413	N≡N	945	H−Br	366
H_2N-NH_2	158	O−H	464	C=O	1077	H−I	298

Fig. B.6 Some bond energies

Bond energies can be used to calculate the approximate value for the **enthalpy change** of reactions.

BOND LENGTH

Bond length is the average distance between the **nuclei** of two **atoms** in a molecule. Bond lengths can be measured by microwave **spectroscopy**. For any two given atoms, the bond gets shorter and stronger as it changes from a **single** to a **double** to a **triple covalent** bond.

BORN–HABER CYCLE

A Born–Haber cycle is an energy cycle which uses **Hess's law** for the calculation of **lattice energies**. The formation of an ionic compound such as sodium chloride is imagined to involve a number of steps whose energy changes are shown against a vertical scale. The arrows of the **endothermic** processes point upwards and those for the **exothermic** processes point downwards – see Fig. B.7.

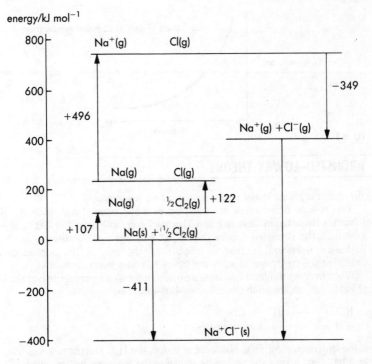

energy/kJ mol^{-1}

Fig. B.7 Born—Haber cycle for sodium chloride

The lattice energy of the **ionic crystal** is calculated as the net result of all the other energy changes in the cycle. For example, the lattice energy of sodium chloride balances the enthalpy of formation ($\triangle H_{f,298}^{\theta} = -411$ kJ mol^{-1}) of NaCl(s), the **atomisation energies** of sodium and chlorine ($\triangle H_{at}^{\theta} = +107$ and $+122$kJ mol^{-1}), the first **ionisation energy** of sodium ($\triangle H_{i}^{\theta} = +496$kJ mol^{1}) and the **electron affinity** of chlorine ($\triangle H_{e}^{\theta} = -349$kJ mol^{-1}), thus:

$$\triangle H_{1}^{\theta} = \triangle H_{f,298}^{\theta}[NaCl] - \triangle H_{at}^{\theta}[Na] - \triangle H_{at}^{\theta}[Cl_2] - \triangle H_{i}^{\theta}[Na] - \triangle H_{e}^{\theta}[Cl]$$
$$= (-411) \quad - (+107) \quad - (+122) \quad - (+496) \quad - (-349)$$
$$= -787\text{kJ mol}^{-1}$$

BOYLE'S LAW

The volume of a fixed mass of gas at a constant temperature is inversely proportional to its pressure: $V \propto 1/p$

If you put your finger over the hole of a bicycle pump and push on the pump handle, you can squeeze the air into a smaller volume and feel the pressure rise. So p gets higher as V gets smaller; see Fig. B.8.
◀ Gas laws, Ideal gas equation ▶

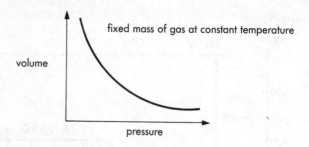

fixed mass of gas at constant temperature

volume

pressure

Fig. B.8

BRONSTED–LOWRY THEORY

An acid is a proton donor and a base is a proton acceptor.

This general definition of **acids** and **bases** was proposed independently by Johannes Bronsted in Denmark and Thomas Lowry in England in 1923 and is known as the Bronsted–Lowry theory. The theory was an advance on the Arrhenius definition because it emphasised the role of the solvent and extended the treatment of acids and bases to non-aqueous systems.

When hydrogen chloride donates a **proton**, it forms its **conjugate base**, the chloride ion, as shown in the following **half-equation**:

$$HCl \longrightarrow H^+ + Cl^-$$
acid $\qquad\qquad$ conjugate base

If the hydrogen chloride is dissolved in water, the H_2O solvent molecules act as a base and accept the proton as shown in the following half-equation:

$$H^+ + H_2O \longrightarrow H_3O^+$$
base $\qquad\qquad$ conjugate acid

Combining the two half-equations gives the complete equation for the **proton transfer** reaction:

$$HCl(aq) + H_2O(l) \longrightarrow H_3O^+(aq) + Cl^-(aq)$$
acid 1 $\qquad$ base 2 $\qquad\qquad$ acid 2 $\qquad$ base 1

The strength of the acid varies with the solvent. In water, HCl, HBr and HI are all equally strong, being completely dissociated into $H_3O^+(aq)$ ions and halide ions. But with liquid ethanoic acid as the solvent, they are not completely dissociated and their strength varies from HI as the strongest to HCl as the weakest acid. Although ethanoic acid, like water, is a polar solvent, it is a weaker base (proton acceptor) than water. By contrast, liquid ammonia is a polar solvent and a stronger proton acceptor than water. Consequently many acids that are only partially dissociated in water are completely dissociated in liquid ammonia.

In water as solvent, the **neutralisation** of a strong acid by a strong base is the converse of the self-ionisation of the solvent:

$$H_3O^+(aq) + OH^-(aq) \rightarrow 2H_2O(l)$$

The comparable equation for a neutralisation in liquid ammonia is:

$$NH_4^+ + NH_2^- \rightarrow 2NH_3$$

Consequently, according to the Bronsted–Lowry theory, in liquid ammonia as solvent, ammonium chloride acts as an acid and 'neutralises' sodium amide acting as a base:

$$NH_4Cl + NaNH_2 \rightarrow NaCl + 2NH_3$$

or simply $NH_4^+ + NH_2^- \rightarrow 2NH_3$

Like the Arrhenius theory, the Bronsted–Lowry theory can provide a quantitative measure of acidity (in terms of $[H^+(aq)]$ and pH) and of acid strength (in terms of the **acid dissociation constant**).

BUFFER SOLUTION

A buffer is a solution of a **weak acid** and its **conjugate base** (or a **weak base** and its **conjugate acid**) whose pH is almost unchanged by the addition of small amounts of acid or alkali. In principle, **strong acids** (or alkalis) could also be regarded as buffers with a very low (or high) pH value. In practice, weak acids (or alkalis) are used to give buffer solutions with pH values near to 7. For example, aqueous sodium dihydrogenphosphate(V) and di-sodium hydrogen-phosphate(V) is a phosphate buffer with a pH of about 7.2 at the maximum buffering point – see Fig. B.9.

acid	$CH_3CO_2H(aq)$	$H_2PO_4^-(aq)$	$NH_4^+(aq)$
conjugate base	$CH_3CO_2^-(aq)$	$HPO_4^{2-}(aq)$	$NH_3(aq)$
pK_a = pH max buffer	4.8	7.2	9.3

Fig. B.9 pH of three buffers

The pH of a buffer solution is governed by the pK_a and the ratio of the **concentrations** of weak acid to conjugate base. For maximum buffering effect, the acid and its conjugate base should have the same concentrations so that the solution can deal equally well with the addition of small amounts of alkali or acid. In a buffer solution of aqueous ethanoic acid and sodium ethanoate:

$$\frac{[CH_3CO_2^-(aq)][H_3O^+(aq)]}{[CH_3CO_2H(aq)]} = K_a = 1.7 \times 10^{-5} \text{ mol dm}^{-3}$$

At the maximum buffering point $[CH_3CO_2^-(aq)] = [CH_3CO_2H(aq)]$, so

$$[H_3O^+(aq)] = K_a = 1.7 \times 10^{-5} \text{ mol dm}^{-3}$$
$$\Rightarrow pH_{(max.buffer)} = pK_a = 4.8$$

In a buffer solution of aqueous ammonia and ammonium chloride:

$$\frac{[NH_4^+(aq)][OH^-(aq)]}{[NH_3(aq)]} = K_b = 1.8 \times 10^{-5} \text{ mol dm}^{-3}$$

At the maximum buffering point $[NH_4^+(aq)] = [NH_3(aq)]$, so

$$[OH^-(aq)] = K_b = 1.8 \times 10^{-5} \text{ mol dm}^{-3}$$
$$\Rightarrow pOH_{(max.buffer)} = pK_b = 4.7$$
$$\Rightarrow pH_{(max.buffer)} \text{ is } pK_w - pK_b = 14 - 4.7 = 9.3$$

A solution of $NH_4^+(aq)$ and $NH_3(aq)$ could be considered as a buffer of the very weakly acidic ammonium ion ($NH_4^+(aq)pK_a = 9.3$) and the ammonia molecule, $NH_3(aq)$, its conjugate base.

If the concentrations of acid and conjugate base are not quite equal, the buffer solution will not be at its maximum buffering point and the pH of the buffer will not equal pK_a.

For a buffer solution at or near its maximum buffering point, the pH can be calculated using the Henderson–Hasselbalch equation:

$$pH_{(buffer)} = pK_a + \log \frac{[\text{conjugate base}]}{[\text{weak acid}]}$$

This is merely an alternative version of the following general equation:

$$\frac{[H_3O^+(aq)][\text{conjugate base}]}{[\text{weak acid}]} = K_a$$

If only small amounts of acid or alkali are added to a buffer solution, the ratio of the concentrations of the acid and its conjugate base hardly changes, so neither does $[H_3O^+(aq)]$. Consequently the logarithm of the ratio [conjugate base]/[weak acid] changes even less, so the pH of the buffer hardly changes at all. However, if large amounts of acid (or alkali) are added in relation to the concentration of the buffer solution, the pH will change. A tiny drop of buffer solution will not cope with a bucket of concentrated acid! The capacity of a buffer is governed by the concentrations of the weak acid and its conjugate base.

CAESIUM CHLORIDE STRUCTURE

The caesium chloride structure is an ionic **crystal structure** which may be described as double simple (interpenetrating) cubic (see Fig. C.1). Each ion has a coordination number of 8, being at the centre of a cube formed by eight oppositely charged ions at the corners of the cube. This crystal structure is adopted by NH_4Cl, NH_4Br, NH_4I, CsCl, CsBr, CsI and CuZn (β-brass). It should not be confused with the **body-centred cubic structure** adopted by pure **alkali metals** in which the **atom** at the centre of the cube is the same as those at the eight corners of the cube.

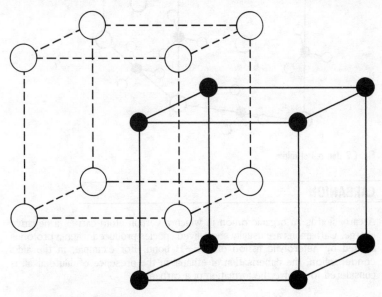

coordination number $Cs^+ = 8$
coordination number $Cl^- = 8$
$\Rightarrow$ ratio of $Cs^+:Cl^-$ is 1:1
hence formula of compound is CsCl

Fig. C.1 Double simple cubic structure of caesium chloride

CALCITE STRUCTURE

The calcite structure is the ionic **crystal structure** adopted by calcium carbonate as its polymorph calcite – see Fig. C.2. The rhombohedral unit cell is like a **face-centred cubic unit cell** of sodium chloride that has been pulled out by diagonally opposite corners. The structure may be seen as spherical calcium ions and planar carbonate ions each occupying the corners and centres of the faces of two interpenetrating rhombohedra. Each ion is surrounded by six oppositely charged ions giving a **coordination number** ratio of 6:6 consistent with the formula $CaCO_3$. The planar carbonate ions make the crystals **anisotropic**.

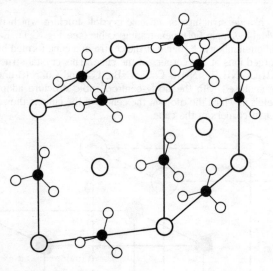

Fig. C.2 Unit cell of calcite

CARBANION

A carbanion is an organic **anion** in which a carbon atom carries a negative charge. Carbanions are usually short-lived species produced when a **proton** is formed by **heterolytic fission** of a C−H bond. For example, in the aldol **condensation**, the dimerisation of ethanal in the presence of dilute alkali is considered to involve the formation of a carbanion:

$$HO:^- \quad + \quad \begin{array}{c} H \ H \\ | \ \ | \\ H-C-C=O \\ | \\ H \end{array} \quad \rightarrow \quad HO:H \quad + \quad \begin{array}{c} H \ H \\ | \ \ | \\ {}^-:C-C=O \\ | \\ H \end{array}$$

ethanal   carbanion

The carbanion acts as a **nucleophile** by attacking the electrophilic carbon in the C=O group of a second ethanal molecule:

$$\begin{array}{c} \underset{\substack{|\\\delta^+\\H}}{\overset{\substack{H \quad H\\|\quad|\,\delta^-}}{H-C-C=O}} + \underset{\substack{|\\H}}{\overset{\substack{H \quad H\\|\quad|}}{\,^-:C-C=O}} \rightarrow \underset{\substack{|\quad|\\H \quad O^-}}{\overset{\substack{H \quad H \quad H \quad H\\|\quad|\quad|\quad|}}{H-C-C--C-C=O}} \overset{H_2O}{\rightarrow} \underset{\substack{|\quad|\quad|\\H \quad OH\,H}}{\overset{\substack{H \quad H \quad H \quad H\\|\quad|\quad|\quad|}}{H-C-C-C-C=O}} + OH^- \end{array}$$

The formation of the carbanion is encouraged by the electronegative oxygen atom and **delocalisation** of the negative charge that would otherwise reside completely on the carbon atom. Carbanions are important transient intermediates in organic reactions but they are less widespread than **carbocations**. The most stable of the simple carbanions is that formed when ethyne is passed into a solution of sodium metal in liquid ammonia: $CH\equiv C^-Na^+$.

CARBOCATION

A carbocation is an organic **cation** in which a carbon atom carries a positive charge. It is formed when a **covalent bond** to a carbon atom breaks heterolytically, with the leaving group (X) taking both electrons:

$$\underset{|}{\overset{|}{-C-X}} \rightarrow \underset{|}{\overset{|}{-C^+}} + :X^-$$

The formation of a carbocation is the **rate-determining step** in the S_N1 hydrolysis of tertiary **halogenoalkanes**. The carbocation is a short-lived and extremely reactive **electrophile**. Carbocations are extremely important transient intermediates in a wide range of organic **reaction mechanisms**.

CARBOHYDRATE

Carbohydrates are a broad class of organic compounds with the general formula $C_xH_{2y}O_y$. The general formula may also be written in the form $C_x(H_2O)_y$ to show that these compounds contain hydrogen and oxygen combined in the same proportion as in water. However, carbohydrates are NOT compounds of carbon and water! They are usually subdivided into mono- and di-saccharides, such as glucose and sucrose, and **polysaccharides**, such as starch and cellulose.

Glucose ($C_6H_{12}O_6$) is of great biochemical importance. In its open chain structure a glucose molecule is seen to possess an **aldehyde group**; hence it is classed as an aldose. It is extremely soluble in water because the OH groups in the molecule can form extensive **hydrogen bonds** with water molecules; see Fig. C.3.

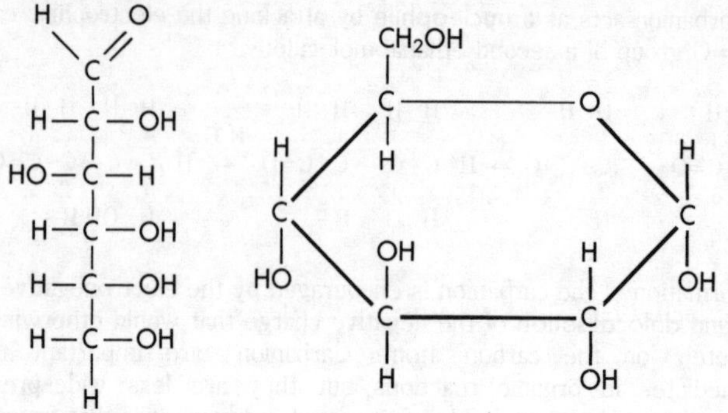

Fig. C.3 D (+) glucose

Sucrose may be regarded as the product of a **condensation reaction** between a glucose and fructose molecule; see Fig. C.4.

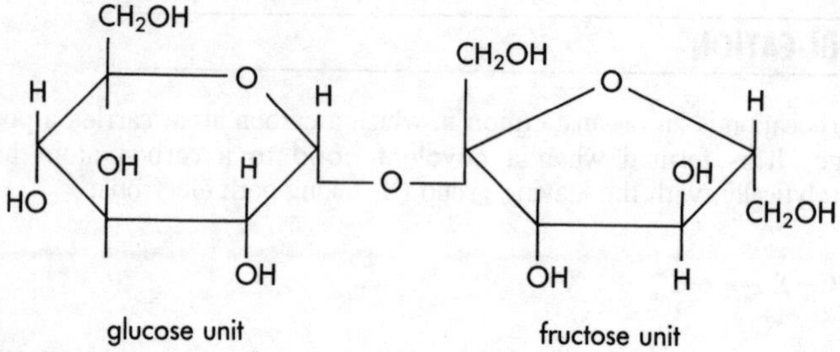

glucose unit   fructose unit

Fig. C.4 Sucrose molecule

Starch is a **condensation polymer** of α-glucose and cellulose is a condensation polymer of β-glucose; see Fig. C.5.

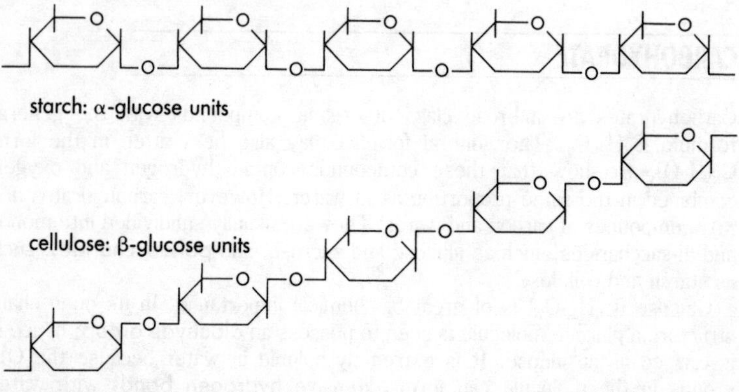

starch: α-glucose units

cellulose: β-glucose units

Fig. C.5 Polysaccharides

CARBONATE

Carbonates are inorganic salts of carbonic acid containing the planar carbonate ion CO_3^{2-}.

Apart from the anomalous lithium carbonate, the **alkali metal** carbonates are soluble in water and are not decomposed when heated in a bunsen flame. The carbonates of other metals are insoluble in water and decompose in a bunsen flame to give the metal **oxide** and carbon dioxide:

$$CaCO_3(s) \xrightarrow{\text{heat}} CaO(s) + CO_2(g)$$

All carbonates give off carbon dioxide with acids such as **ethanoic** acid, a **carboxylic** acid that is stronger than carbonic acid, but not with acids such as **phenol**, an acid that is weaker than carbonic acid:

$$2H^+(aq) + CO_3^{2-}(aq) \rightarrow H_2CO_3(aq) \rightarrow H_2O(1) + CO_2(g)$$

Calcium carbonate is mined as limestone and is of considerable industrial importance.

CARBONYL COMPOUNDS

Inorganic carbonyl compounds are neutral **complexes** formed by the reaction of **transition metals** with carbon monoxide. The **oxidation state** of the metal remains at zero and its **electron configuration** is increased to that of a **noble gas** by **dative bonding** of the :CO ligands. Cobalt, for example, achieves the electronic configuration of krypton by forming the following carbonyl: $(CO)_4Co-Co(CO)_4$. The recovery and purification of metallic nickel depends upon the formation and decomposition of tetracarbonylnickel(0).

Organic carbonyl compounds are **aldehydes** and **ketones** whose structures contain the carbonyl **functional group** $>C=O$. These two classes of compound are isomeric and show both similarities and differences in properties. In aldehydes the carbonyl group is always attached to one hydrogen atom to give the functional group $-CHO$. In ketones the carbonyl group is always attached to two carbon atoms. The reactions of the $>C=O$ group may be compared to the reactions of the $>C=C<$ group.

▶ PHYSICAL PROPERTIES

Aldehydes and ketones are typical covalent substances whose molecules are made polar by the $>C=O$ group. Permanent **dipole–dipole attractions** contribute to the intermolecular forces which make their melting and boiling points higher than those of **aliphatic hydrocarbons** and **ethers** but lower than those of aliphatic **alcohols** of similar **molar mass** (see Fig. C.6). Only methanal and ethanal are gases at 25°C and 1 atm. The other aldehydes and the ketones met in A-level chemistry are liquids. The lower ketones are good **polar solvents** for many organic reactions because they will dissolve both ionic

Name	M_r	$T_b/°C$	Dipole moment/d	Structure	Intermolecular forces
butane*	58	0	0	$CH_3(CH_2)_2CH_3$	van der Waals
methoxyethane	60	11	1.23	$C_2H_5OCH_3$	dipole-dipole
propanal	58	49	2.52	CH_3CH_2CHO	dipole-dipole
propanone	58	56	2.88	$(CH_3)_2CO$	dipole-dipole
propan-1-ol	60	98	1.68	$CH_3CH_2CH_2OH$	dipole-dipole and H-bonding
propan-2-ol	60	83	1.66	$(CH_3)_2CHOH$	dipole-dipole and H-bonding

* non-polar molecule included for comparison with polar molecules

Fig. C.6 Physical properties of some polar compounds

and covalent substances. Carbonyl compounds do not hydrogen-bond to each other because their molecules do not have a hydrogen atom attached to a sufficiently electronegative atom (such as N, O or F). Consequently **nucleophilic substitution** reactions occur readily in, for example, propanone as solvent (and faster than in water or alcohols as solvents) because the **nucleophiles** do not become **solvated** by hydrogen-bonding to the propanone molecules. Carbonyl compound molecules will form **hydrogen bonds** with water molecules, so the lower aldehydes and ketones are soluble in water.

▶ CHEMICAL PROPERTIES

The lower members are extremely flammable and their vapours ignite explosively, so great care is needed when handling propanone (acetone). It must be kept away from naked flames. Aldehydes are readily oxidised to **carboxylic acids** by warming with acidified aqueous sodium dichromate(VI) or with **Fehling's solution**. By contrast, most ketones are not readily oxidised to carboxylic acids but some **cyclic** ketones will degrade to a mixture of carboxylic acids (with fewer carbon atoms than the original ketone) when refluxed with vigorous **oxidising agents** such as acidified potassium manganate(VII).

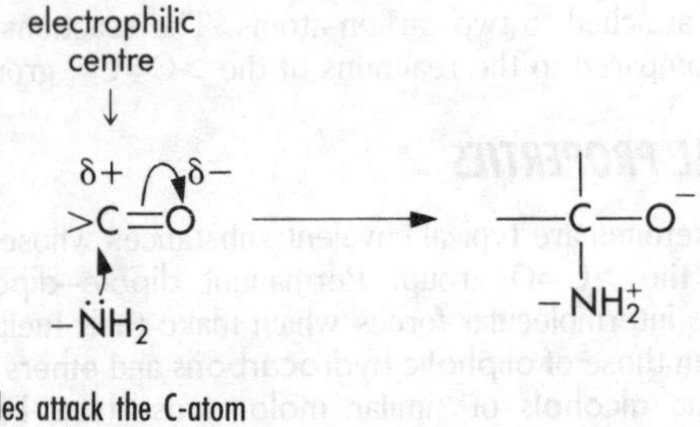

Fig. C.7 Nucleophiles attack the C-atom

Oxygen is more electronegative than carbon so the C=O bond is polar and the C-atom is an electrophilic centre open to attack by nucleophiles (see Fig.

C.7). Moreover, the C=O bond is polarisable and the O-atom can accept a displaced π-electron pair to form a fairly stable intermediate **oxoanion** and make the α-carbon atom a nucleophilic centre open to attack by **electrophiles** (see Fig. C.8).

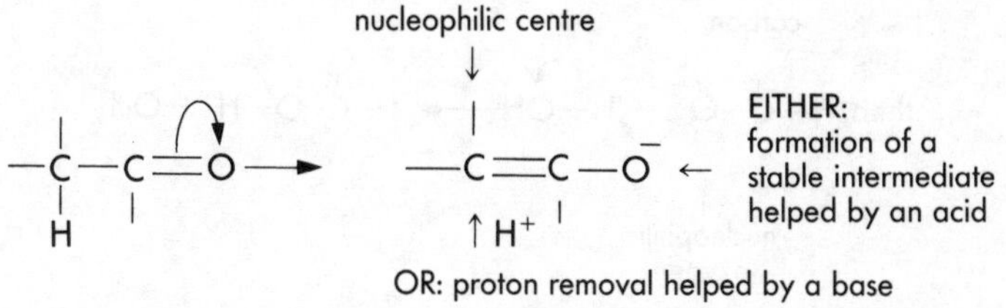

nucleophilic centre
↓

EITHER:
formation of a
stable intermediate
helped by an acid

OR: proton removal helped by a base

You should contrast this with the C = C bond in which a C-atom is attacked initially by electrophiles. If nucleophiles were to attack the C = C bond initially, the intermediate carbanion would be relatively unstable.

Fig. C.8 Electrophiles attack the α-C-atom

A hydrogen atom on the α-carbon atom (the carbon atom next to the C=O group) is readily replaced by a **halogen atom**. So, **aqueous** propanone and iodine readily undergo an acid-catalysed reaction:

$$CH_3-\underset{\underset{O}{\|}}{C}-CH_3 + I_2(aq) \xrightarrow[OH^-(aq)]{H^+(aq) \text{ or}} CH_3-\underset{\underset{O}{\|}}{C}-CH_2I + H^+(aq) + I^-(aq)$$

Ethanal and methyl ketones give a positive result (yellow precipitate of CHI_3) with the iodoform test because their structures contain the $CH_3-\underset{|}{C}=O$ group.

The oxygen in the carbonyl group may be replaced by chlorine by reaction with phosphorus pentachloride under anhydrous conditions; for example:

$$(CH_3)_2CO \xrightarrow[\text{conditions}]{PCl_5 \text{ in anhydrous}} (CH_3)_2CCl_2 \quad \text{2,2-dichloropropane}$$

Carbonyl compounds are reduced to alcohols, aldehydes giving **primary alcohols** and ketones giving **secondary alcohols**. The **reduction** of carbonyl compounds by sodium tetrahydridoborate or lithium tetrahydridoaluminate may be seen as the **addition** of hydrogen across the C=O double bond, with the BH_4^- and AlH_4^- ions supplying hydride anions, $:H^-$, for nucleophilic attack upon the electrophilic carbon atom in the carbonyl group and the water molecules supplying hydrogen cations for electrophilic attack upon the nucleophilic oxygen atom in the carbonyl group (see Fig. C.9).

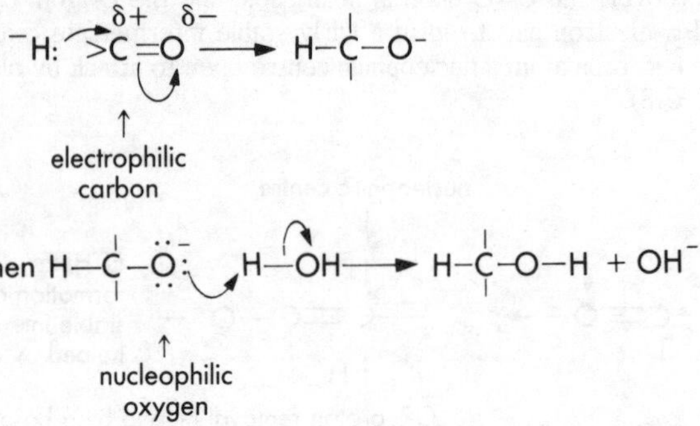

$$-H: \overset{\delta+}{\underset{}{>}}C = \overset{\delta-}{O} \longrightarrow H - \overset{|}{\underset{|}{C}} - O^-$$

↑
electrophilic
carbon

$$\text{then } H - \overset{|}{\underset{|}{C}} - \overset{\cdot\cdot}{\underset{\cdot\cdot}{O}}: \quad H - OH \longrightarrow H - \overset{|}{\underset{|}{C}} - O - H + OH^-$$

↑
nucleophilic
oxygen

Fig. C.9 Reduction of carbonyl compounds by metal hydrides

Carbonyl compounds undergo **nucleophilic addition reactions**. For example, the addition of hydrogen cyanide to propanone is the first stage of a multi-stage process for the manufacture of poly(methyl 2-methylpropenoate) and other acrylic **polymers**. Sodium hydroxysulphonate **crystals** are formed when carbonyl compounds are shaken with freshly prepared aqueous sodium hydrogensulphite – see Fig. A.16. Recrystallisation of this addition compound, followed by regeneration of the original carbonyl compound by adding acid (or alkali) to the crystals, provides a way of purifying liquid (or gaseous) carbonyl compounds.

Nucleophile	Reagents and conditions	Formula and names of addition products	
$: H^-$	$NaBH_4(aq)$ or $LiAlH_4$ in dry ethoxyethane then add water	$R' - \overset{OH}{\underset{R''}{\overset{\|}{\underset{\|}{C}}}} - H$	If $R' = CH_3$ & $R'' = H$ then ethanol. If $R' = R'' = CH_3$ then propan-2-ol
$: CN^-$	aqueous sodium cyanide (KCN) then excess mineral acid	$R' - \overset{OH}{\underset{R''}{\overset{\|}{\underset{\|}{C}}}} - CN$	If $R' = CH_3$ & $R'' = H$ then 2-hydroxy-propanenitrile. If $R' = R'' = CH_3$ then 2-hydroxy-2-methylpropane-nitrile
$: SO_3^-$	freshly made saturated aqueous sodium hydrogensulphite	$R' - \overset{OH}{\underset{R''}{\overset{\|}{\underset{\|}{C}}}} - SO_3^- Na^+$	If $R' = CH_3$ & $R'' = H$ then sodium 2-hydroxy-propane sulphonate
$: C_2H_5^-$	Grignard reagent $CH_3CH_2 - Mg - Br$ in dry ethoxyethane then add water	$R' - \overset{OH}{\underset{R''}{\overset{\|}{\underset{\|}{C}}}} - CH_2CH_3$	If $R' = CH_3$ & $R'' = H$ then butan-2-ol. If $R' = R'' = CH_3$ then 2-methylbutan-2-ol

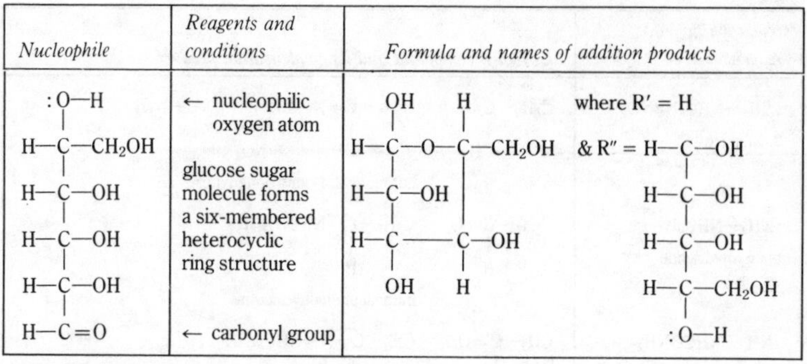

Nucleophile	Reagents and conditions	Formula and names of addition products		
$:O—H$ $\|$ $H—C—CH_2OH$ $\|$ $H—C—OH$ $\|$ $H—C—OH$ $\|$ $H—C—OH$ $\|$ $H—C=O$	← nucleophilic oxygen atom glucose sugar molecule forms a six-membered heterocyclic ring structure ← carbonyl group	$\underset{\|}{O}H \quad H$ $H—C—O—C—CH_2OH$ $\|$ $H—C—OH$ $\|$ $H—C———C—OH$ $\|$ $OH \quad H$	where $R' = H$ & $R'' = H—C—OH$ $\|$ $H—C—OH$ $\|$ $H—C—OH$ $\|$ $H—C—CH_2OH$ $\|$ $:O—H$	

Fig. C.10 Addition reactions of carbonyl compounds

Nucleophilic addition reactions of aldehydes and ketones are summarised in Fig. C.10. Compounds containing $—NH_2$ undergo a nucleophilic addition between the N-atom and the electrophilic C-atom in the $>C=O$ but the product loses water molecules by an elimination reaction to form a $N=C$ **double bond**; see Fig. C.11.

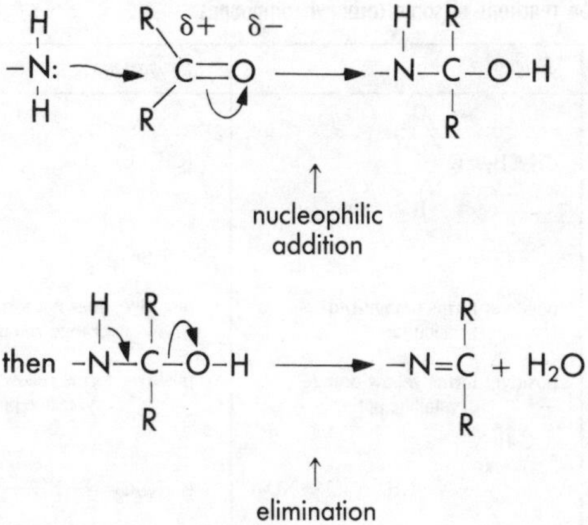

nucleophilic
addition

then $-N-C-O-H \longrightarrow -N=C + H_2O$

elimination

Fig. C.11 Condensation (addition–elimination) reaction

The **condensation** (addition–elimination) **reactions** of some carbonyl compounds are summarised in Fig. C.12. The reactions with ammonia are often complicated, but with compounds derived from ammonia the reactions usually give simple, well-defined condensation products which can be recrystallised and used to identify the carbonyl compound or to distinguish between aldehydes and ketones; see Fig. C.13.

Nucleophilic reagent	Carbonyl compound and its condensation product
$:NH_2—NHC_6H_4(NO_2)_2$ 2,4-dinitro-phenylhydrazine	$\underset{\underset{CH_3}{\vert}}{C_2H_5—C}=O$ $\quad$ $\underset{\underset{CH_3}{\vert}}{C_2H_5—C}=N—NH—\bigcirc—NO_2$ $\underset{NO_2}{}$ butanone 2,4-dinitrophenyldrazone
$:NH_2—NHC_6H_5$ phenylhydrazine	$\underset{\underset{H}{\vert}}{C_3H_7—C}=O$ $\quad$ $\underset{\underset{H}{\vert}}{C_3H_7—C}=N—NHC_6H_5$ butanal phenylhydrazone
$:NH_2—NHCONH_2$ semicarbazide	$\underset{\underset{CH_3}{\vert}}{CH_3—C}=O$ $\quad$ $\underset{\underset{CH_3}{\vert}}{CH_3—C}=N—NHCONH_2$ propanone semicarbazone
$:NH_2OH$ hydroxylamine	$\underset{\underset{H}{\vert}}{CH_3—C}=O$ $\quad$ $\underset{\underset{H}{\vert}}{CH_3—C}=N—OH$ ethanal oxime
$:NH_3$ ammonia	$\underset{\underset{H}{\vert}}{H—C}=O$ $\quad$ $(CH_2)_6N_4$ heterocyclic fused rings hexamethylenetetramine 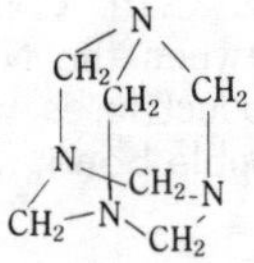

Fig. C.12 Condensation reactions of some carbonyl compounds

	Propanal	Propanone
formula	$CH_3CH_2—\overset{\displaystyle O}{\underset{\displaystyle H}{C}}$	$(CH_3)_2C=O$
boiling point/°C	49	56
reaction with Fehling's solution	positive: forms orange-red precipitate	negative: does not form an orange-red ppt.
reaction with Brady's reagent	positive: forms yellow-orange crystalline ppt.	positive; forms yellow-orange crystalline ppt.
formula of the yellow-orange crystalline ppt.	$\underset{\underset{H}{\vert}}{\overset{C_2H_5}{\vert}}C=N—NH—\bigcirc—NO_2$ $\underset{NO_2}{}$	$(CH_3)_2C=N—NH—\bigcirc—NO_2$ $\underset{NO_2}{}$
melting point/°C	156	128

Fig. C.13 Distinguishing an aldehyde from a ketone

Carbonyl compounds are manufactured by controlled oxidation of alcohols or **hydrocarbons, hydration** of **alkenes** by the Wacker process and as co-products of other processes such as the cumene process. Industrial uses of carbonyl compounds include solvents and the production of **plastics** such as bakelite, formica, melamine and **perspex,** and epoxy and poly(carbonate) resins.

CARBOXYLIC ACIDS

Carboxylic acids are organic compounds containing $-C\overset{\displaystyle O}{\underset{\displaystyle OH}{}}$ the carboxyl **functional group** in their molecular structure. Although the carboxyl group seems to consist of an OH and a C=O group, carboxylic acids show few similarities to **alcohols, aldehydes** and **ketones**.

▶ NOMENCLATURE

The names of carboxylic acids end in **–anoic** acid if one carboxyl group is part of the hydrocarbon chain and **–anedioic** acid if two carboxylic acids are part of the carbon chain. **Aromatic** acids, with the C-atom of the $-CO_2H$ attached to but not part of the benzene ring, end with the words -carboxylic acid. The traditional names for some acids are still widely used. Ethanoic acid (acetic acid) and benzenecarboxylic acid (benzoic acid) are typical carboxylic acids.

▶ PHYSICAL PROPERTIES

The physical properties are those expected of **polar** covalent molecular structures capable of **hydrogen bonding**. Their melting and boiling points are higher than those of **aliphatic** and **aromatic hydrocarbons** of similar molar mass but not as high as you might expect because the carboxylic acid molecules hydrogen-bond together in pairs, called **dimers** (see Fig. C.14).

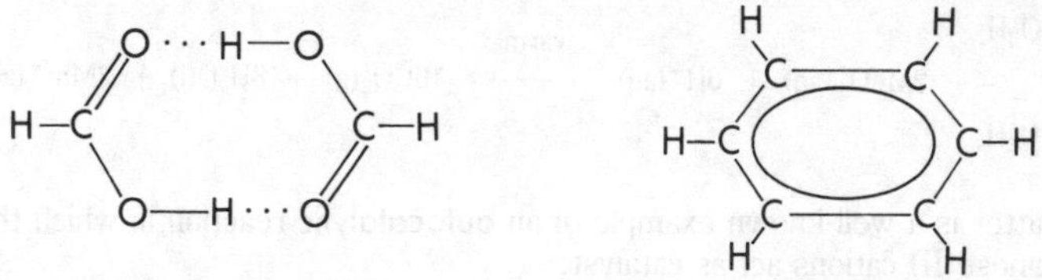

dimer of methanoic acid
m.pt./°C=8.3
b.pt./°C=100.5
M_r(dimer) = 92

benzene
m.pt./°C=5.4
b.pt./°C=80.0
M_r(dimer) = 78

Fig. C.14 Comparison of methanoic acid and benzene

These dimers are non-polar and not hydrogen-bonded to each other but because of their symmetrical shape they can pack tightly together to give plate-like crystals with fairly high melting points. This explains the unexpected similarity in the physical properties of methanoic acid and benzene (Fig. C.14). The lower fatty acids have unpleasant smells in contrast to the pleasant fruity smells of their **esters** – see Fig. C.15.

Ethanoic acid mixes with water in all proportions. Benzoic acid is soluble only in hot water and crystallises readily when the solution is cooled.

CARBOXYLIC ACIDS

Systematic name	Traditional name	Latin origin	Smell
ethanoic acid	acetic acid	acetum: vinegar	vinegar
butanoic acid	butyric acid	butyrum: butter	rancid butter
hexanoic acid	caproic acid	caper: goat	old goat

Fig. C.15 Smells of some carboxylic acids

▶ CHEMICAL PROPERTIES

Carboxylic acids are the **oxidation** products of **primary alcohols** and **aldehydes** and are not readily oxidised, but they will burn in excess air or oxygen to form carbon dioxide and water. Benzoic acid can be used to calibrate a **bomb calorimeter**:

$$C_6H_5CO_2H(s) + O_2(g) \rightarrow 6CO_2(g) + 3H_2O(l); \; \triangle H_c^{\theta} = -3227 \text{ kJ mol}^{-1}$$

Methanoic (formic) acid and ethanedioic (oxalic) acid are exceptional in reacting with oxidising agents. Methanoic acid, whose structure seems to contain an aldehyde group $-CHO$, reduces **Fehling's solution**:

$$H-C\overset{\displaystyle O}{\underset{\displaystyle OH}{\big<}} + 2Cu^{2+} + 4OH^- \rightarrow CO_2 + 3H_2O + Cu_2O$$

and warm aqueous ethanedioic acid can be **titrated** with acidified aqueous potassium manganate(VII) solution:

$$5 \begin{vmatrix} CO_2H \\ \\ CO_2H \end{vmatrix} + 2MnO_4^-(aq) + 6H^+(aq) \xrightarrow{\text{warm}} 10CO_2(g) + 8H_2O(l) + 2Mn^{2+}(aq)$$

The latter is a well-known example of an **autocatalytic** reaction in which the manganese(II) cations act as catalyst.

The alkyl chain of **aliphatic** carboxylic acids undergoes **free-radical** chain substitution reactions when chlorine is bubbled into the boiling acid in UV light. The reaction can be monitored by measuring the increase in mass of the liquid; for instance:

$$CH_3CO_2H \rightarrow CH_2ClCO_2H \rightarrow CHCl_2CO_2H \rightarrow CCl_3CO_2H.$$

The $-OH$ in the carboxyl group is replaced by $-Cl$, by reaction with phosphorus chlorides or sulphur dichloride oxide, to form the acyl chloride and give off steamy fumes of hydrogen chloride:

$$CH_3-C\overset{\displaystyle O}{\underset{\displaystyle OH}{\big/\!\big/}} + PCl_5 \rightarrow CH_3-C\overset{\displaystyle O}{\underset{\displaystyle Cl}{\big/\!\big/}} + PCl_3O + HCl$$

ethanoyl	steamy
chloride	fumes

Carboxylic acids are reduced to primary alcohols by a powerful reducing agent such as $LiAlH_4$ in dry ethoxyethane. The method CANNOT be used to prepare aldehydes because they are reduced to primary alcohols even more readily than the carboxylic acids are.

Name	Formula	pK_a
trichloroethanoic acid	CCl_3CO_2H	0.7
dichloroethanoic acid	$CHCl_2CO_2H$	1.3
chloroethanoic acid	CH_2ClCO_2H	2.9
benzoic acid	$C_6H_5CO_2H$	4.2
ethanoic acid	CH_3CO_2H	4.8
propanoic acid	$C_2H_5CO_2H$	4.9

Fig. C.16 Strengths of some carboxylic acids

Most carboxylic acids are **weak acids**; in aqueous solution they partially ionise into hydrogen cations and carboxylate anions. The strength of a carboxylic acid is affected by its molecular structure. Electron-attracting groups attached to the carboxyl group help further to delocalise the negative charge, stabilise the carboxylate anion and make the carboxylic acid stronger. Trichloroethanoic acid could be regarded as a **strong acid!** Electron-donating groups have the opposite effect – see Fig. C.16. Although ethanoic acid is a weak acid ($pK_a = 4.8$) it is stronger than carbonic acid ($pK_a = 6.4$) and **phenol** ($pK_a = 9.9$) because **delocalisation** of the negative charge makes the ethanoate anion more stable than the hydrogencarbonate anion and the phenoxide anion:

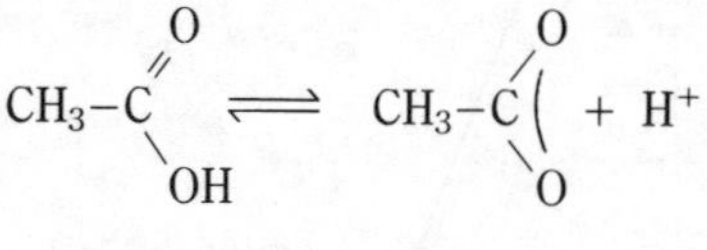

$$CH_3-C\underset{OH}{\overset{O}{\big\backslash\!\!\big/}} \rightleftharpoons CH_3-C\underset{O}{\overset{O}{\big\backslash\!\!\big/}} + H^+$$

Consequently, most carboxylic acids can be distinguished from phenol because they displace CO_2 from **carbonates** and hydrogencarbonates but phenol does not:

$$CH_3CO_2H(aq) + CO_3^{2-}(aq) \rightarrow CH_3CO_2^-(aq) + H_2O(l) + CO_2(g)$$

Carboxylic acids react with calcium, sodium, magnesium and some other active metals to produce hydrogen gas:

$$CH_3CO_2H(aq) + Na(s) \rightarrow CH_3CO_2^-(aq) + Na^+(aq) + ½H_2(g)$$

These reactions are **redox reactions** and not simply **acid–base** (or proton-transfer) reactions because the metal atom is oxidised to the cation and the hydrogen (cation) is reduced to the hydrogen molecule. However, the partially ionised carboxylic acid is merely involved in transfer of protons for **reduction** by the metal. Carboxylic acids dissolve in aqueous sodium hydroxide to form solutions of their sodium salts. They also react with basic metal **oxides** and with ammonia to form salts and water or just salts.

Carboxylic acids react with **alcohols** and an acid **catalyst** (HCl or H_2SO_4) by direct esterification to form esters and water. Carboxylic acids can be prepared by the oxidation of primary alcohols and by the hydrolysis of esters, amides and nitriles.

CATALYSIS

The use of catalysts to speed up chemical processes.

CATALYST

A catalyst speeds up a reaction without being consumed by it, so the catalyst does not appear as a reactant in the overal chemical equation. Homogeneous and heterogeneous catalysts provide alternative reaction pathways with activation energies that are lower than those of the uncatalysed reactions (see Fig. C.17). In a reversible reaction, the catalyst speeds up the rate of attainment of equilibrium without altering the equilibrium composition. It does this by lowering the activation energies of the forward and reverse reactions by the same extent. If a catalyst is one of the products of the chemical process, the reaction is autocatalytic.

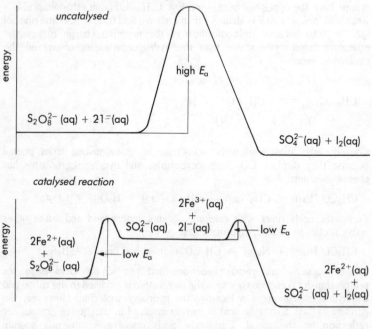

Fig. C.17 Alternative reaction pathway with catalyst

Enzymes are extremely efficient biochemical catalysts whose highly specific action depends upon the presence of **active sites** in the complex tertiary structure of the molecules. **d–block elements** and their compounds are a major source of industrially important inorganic catalysts whose action depends upon the availability of unfilled d-**orbitals** in their atoms or ions.

CATHODE

The cathode is the negatively charged **electrode** in an **electrolytic** cell. In industrial electrolytic refining of metals such as copper, the cathode consists of the pure metal. In industrial electroplating, the cathode is the metal article to be plated. The production of chlorine, sodium hydroxide and hydrogen by the electrolysis of brine employs a flowing mercury cathode and a precise controlled voltage that prevents the discharge of hydrogen ions and permits the discharge of sodium ions into the mercury:

$$Na^+(aq) + e^- \xrightarrow[\text{cathode}]{\text{mercury}} Na(\text{Hg-amalgam})$$

In general, however, electrolysis of aqueous salts of metals above hydrogen in the electrochemical series will result in the discharge of **hydrogen ions** at the cathode and the solution around the cathode becoming alkaline as the **concentration** of hydroxide ions increases:

$$2H^+(aq) + 2e^- \xrightarrow[\text{cathode}]{\text{at the}} H_2(g)$$

CATION

A cation is a positively charged **atom** or group of atoms. The charge on a simple cation may be related to the **electronic configuration** of the atom and the position of the element in the **periodic table**. Simple cations carrying a charge of $1+$, $2+$ and $3+$ are formed by metallic elements in Groups I, II and III respectively, and by various **transition metals**. The **polarising power** of a cation increases as the charge density increases – that is, with increasing charge and decreasing size of the ion.

The hydronium ion, $H_3O^+(aq)$, ammonium ion, NH_4^+, and **amino acids** in highly acidic solutions are examples of cations composed of non-metallic elements. Many aqueous cations, especially those of the transition metals, are **complex cations** in which water molecules are bound as **ligands** by **dative bonds** to the central ion.

During **electrolysis**, ions such as the hydronium ion, $H_3O^+(aq)$, and cations of the noble metals such as Cu^{2+}, move towards the negative **electrode** (cathode), where they may be discharged, hydrogen being liberated as a gas and the metals being deposited on the surface of the electrode.

◀ Hydrogen ion ▶

CELL DIAGRAM

apparatus diagrams

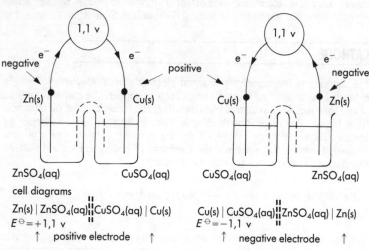

cell diagrams

$Zn(s) \mid ZnSO_4(aq) \parallel CuSO_4(aq) \mid Cu(s)$
$E^{\ominus} = +1,1 \text{ v}$
↑ positive electrode ↑

$Cu(s) \mid CuSO_4(aq) \parallel ZnSO_4(aq) \mid Zn(s)$
$E^{\ominus} = -1,1 \text{ v}$
↑ negative electrode ↑

Fig. C.18 Distinction between apparatus and cell diagrams

A cell diagram is a diagram to represent an **electrochemical cell** and the reaction which takes place in it when a current is allowed to flow. A cell diagram should not be confused with a diagram of the actual *physical apparatus* used to construct an electrochemical cell (see Fig. C.18). The sign given to the e.m.f. associated with the cell diagram of an electrochemical cell refers to the diagram's right-hand **half-cell**.

The following cell diagram illustrates the international convention adopted in writing cell diagrams:

$$Pt \mid H_2(g) \mid H^+(aq) \vdots \vdots [MnO_4^-(aq) + 8H^+(aq)],[Mn^{2+}(aq) + 4H_2O(l)] \mid Pt$$
$\uparrow$ $\uparrow$ $\uparrow$ $\uparrow$
0 +1 +7 +2 [oxidation states]

The vertical solid lines represent a phase boundary, for example, between the solid platinum metal **electrode** and the hydrogen gas or the aqueous solution of manganese compounds and between the hydrogen gas and the aqueous hydrogen ions. The vertical broken lines represent a junction between two liquids. A pair of parallel broken lines indicate that any **liquid junction potential** has been eliminated and may be discounted when calculating the e.m.f. of the electrochemical cell. The species with the lower **oxidation states** are written nearest to the electrodes and those with the higher oxidation states are written nearest to the liquid junction.

CHAIN REACTION

A chain reaction is a reaction that proceeds by a cycle of steps in which a product from one step initiates a subsequent step in the cycle.

Nuclear chain reactions often involve **neutron** capture by an **atom** which then disintegrates to yield two neutrons, each of which could be captured by other atoms to cause disintegration and the production of more neutrons. In a nuclear reactor, a moderator and control rods may be used to absorb the neutrons and prevent the nuclear chain reaction becoming critical and producing an explosion.

Chemical chain reactions are usually gas phase reactions involving **free radicals**. An inorganic chain reaction occurs when the explosion of a mixture of hydrogen and chlorine is triggered by light from a flashbulb. Chlorine molecules split into free radicals: $Cl_2 \rightarrow 2Cl\cdot$ which attack hydrogen molecules to produce hydrogen chloride and $H\cdot$ free radicals to attack chlorine molecules and produce more $Cl\cdot$ radicals:

$$Cl\cdot + H_2 \rightarrow HCl + H\cdot \text{ then } H\cdot + Cl_2 \rightarrow HCl + Cl\cdot$$

The light-initiated bromination of methane is the simplest example of an organic **substitution reaction** involving a (free) radical chain mechanism. The chain reaction involves three steps.

Initiation step

Two free radicals are formed by **homolytic fission** of the $Br-Br$ **covalent bond** when the bromine **molecule** absorbs light:

$$Br-Br \xrightarrow{\text{light}} 2Br\cdot$$

Propagation step

First a bromine radical attacks a methane molecule to produce a methyl radical and a hydrogen bromide molecule, and second a methyl radical attacks a bromine molecule to produce a bromomethane molecule and regenerate a bromine radical:

$$Br\cdot + H-CH_3 \rightarrow H-Br + \cdot CH_3$$
$$\text{then } Br-Br + \cdot CH_3 \rightarrow Br-CH_3 + \cdot Br$$

Termination step

The radicals are removed by, for example, two methyl radicals forming an ethane molecule, a bromine radical and methyl radical forming a bromomethane molecule or even two bromine radicals reforming a bromine molecule:

$$\cdot CH_3 + \cdot CH_3 \rightarrow CH_3-CH_3$$
$$\text{or } Br\cdot + \cdot CH_3 \rightarrow CH_3Br \text{ or } 2Br\cdot \rightarrow Br_2$$

Notice that the propagation steps form a radical chain in which each step in the chain provides the radical for the next step in the chain and that the net effect of the propagation steps is the overall reaction of the bromine and methane to form bromomethane and hydrogen bromide:

$Br_2 + CH_4 \rightarrow HBr + CH_3Br$

The bromination of methane actually produces four different substitution products (CH_3Br, CH_2Br_2, $CHBr_3$ and CBr_4) in varying amounts depending upon the conditions because bromomethane molecules can participate in the propagation steps – for example:

$Br\cdot + H{-}CH_2Br \rightarrow H{-}Br + \cdot CH_2Br$
then $Br{-}Br + \cdot CH_2Br \rightarrow CH_2Br_2 + H{-}Br$

Similarly, dibromomethane molecules could form tribromomethane molecules which could then form tetrabromomethane molecules.

The substitution of hydrogen by chlorine in the methyl group (but not the benzene ring) of methylbenzene (toluene) is a homolytic free radical chain reaction which gives a mixture of products:

$C_6H_5{-}CH_3 \rightarrow C_6H_5{-}CH_2Cl \rightarrow C_6H_5{-}CHCl_2 \rightarrow C_6H_5{-}CCl_3$

The most complicated chain reaction is probably the combustion of alkanes in air.

CHARLES' LAW

At constant pressure, the volume of a fixed mass of gas is directly proportional to its temperature on the Kelvin scale: $V \propto T$

To remember this law, it helps to think of the air in a hot air balloon being heated and the balloon rising. V gets bigger as T gets higher; see Fig. C.19.

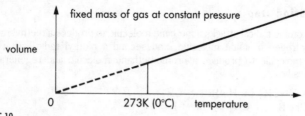

Fig. C.19

◀ Gas laws, Ideal gas equation ▶

CHELATE COMPLEX

A chelate **complex** is a compound containing a complex ion involving polydentate ligands. The word chelate comes from a Greek word meaning a claw. When, for example, the **bidentate ligand** 1,2-diaminoethane forms a chelate complex with chromium, the two carbon atoms, the two nitrogen atoms and the **transition** metal ion form a five-membered ring and the chromium atom seems to be held in a claw:

$$
\begin{array}{cc}
\searrow \text{Cr} \swarrow \\
H_2N \quad NH_2 \\
| \qquad | \\
H_2C - CH_2
\end{array}
$$

Ethylenediaminetetraacetic acid (**EDTA**) is a powerful chelating agent which forms complexes with most metal **cations** and is used in the determination of calcium and magnesium ions in hard water. Chelate complexes are formed by many sequestering agents, such as polyphosphates used in **detergents** for water softening by removing Ca^{2+} and Mg^{2+} ions.

CHEMICAL EQUILIBRIUM

Chemical equilibrium is a dynamic molecular state in which the intensive properties of the reactants and products in a **reversible reaction** have become constant and remain constant with time.

When a reversible reaction is at equilibrium, the rate of the forward reaction is equal to the rate of the reverse reaction and the composition of the system can be described by the **equilibrium constant** value. For example, at about 450°C and 200 atm, the reversible reaction of nitrogen and hydrogen to form ammonia can form an equilibrium mixture containing about 10 per cent ammonia. An iron **catalyst** can accelerate the attainment of this equilibrium state but it cannot alter the composition of the mixture.

◀ Equilibrium ▶

CHIRALITY

Chirality is a term used to refer to molecules which do not have a plane of symmetry and which therefore possess a structure which cannot be superimposed upon its mirror image. If the **molecule** contains a carbon atom to which four different atoms or groups of atoms are attached, that carbon atom is called an *asymmetric* carbon atom and is a *chiral centre* in the molecule. Molecules of α-**amino acids** except glycine have one chiral centre. Monosaccharides such as glucose and fructose have several chiral centres. Chirality usually gives rise to optical activity, and the mirror image **isomers** of the molecules are called **enantiomers**.

◀ Optical activity ▶

CHLORIDES

Chlorides are binary compounds of chlorine with metallic and non-metallic elements.

The **s-block metals** form ionic chlorides which have **giant structures** and dissolve in water to form **neutral solutions**. Many of these chlorides occur naturally. The UK chlor-alkali industry depends upon the salt deposits found in Cheshire and elsewhere. By contrast, the **p-block** non-metals such as silicon, phosphorus, sulphur) form covalent molecular chlorides that hydrolyse in water to give acidic solutions. These do not occur naturally but are produced by reaction of the non-metal with chlorine – for instance:

$$P_4(s) + 10Cl_2(g) \rightarrow 4PCl_5(l); \triangle H^0 = -1774 \text{ kJ mol}^{-1}$$

When a metallic element able to form more than one chloride burns in chlorine, the compound formed is often covalent with the metal in its higher **oxidation state** – for example:

$$2Fe(s) + 3Cl_2(g) \rightarrow 2FeCl_3(s) \text{ [iron(III) chloride]}$$

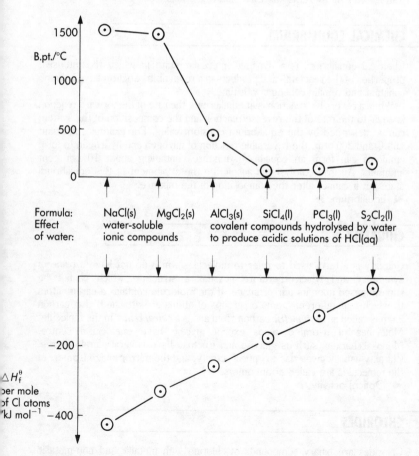

Fig. C.20 Trend in properties of chlorides

The tendency to hydrolyse increases as the chlorides becomes more covalent in character (see Fig. C.20). The pattern in the structure and bonding of the chlorides is one of increasing covalent character with increasing **atomic number** of the elements from left to right across a period of the **periodic table** and of increasing ionic character with increasing atomic number of the elements from top to bottom down a group in the periodic table (see Fig. C.21). This pattern can be discussed in terms of the **polarising power** of the cations and the **polarisability** of the chloride anion.

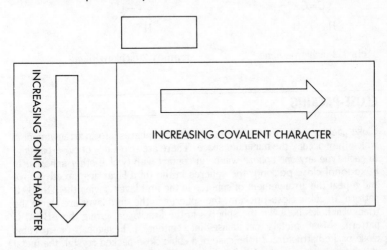

Fig. C.21 Patterns in covalent—ionic character of the oxides, chlorides and hydrides of the elements

Concentrated hydrochloric acid is used in the flame test for metals because the resulting metal chlorides are usually more volatile than the other inorganic compounds of the metals.

CHROMATOGRAPHY

Chromatography is a term which describes techniques for separating mixtures of substances. The principle involved is that a mobile phase (such as a liquid) travels over or through a stationary phase (such as a solid) and that separation of the components of the mixture depends upon their differing interactions with the two phases. The first chromatographic technique was developed in 1906 by Mikhail Tswett, a Russian botanist, for separating plant pigments. He put a solution of the pigments into a column packed with aluminium oxide and washed the solution through the column with more solvent. The pigments separated into coloured bands (Greek *chromos* – colour) which could be individually collected as they left the bottom of the column.

Chromatography now includes techniques such as **gas chromatography**, **paper chromatography** and **thin-layer chromatography**.

CIS-TRANS ISOMERISM

Cis-trans isomerism is another name for a special form of **stereoisomerism** called **geometric(al) isomerism**. The isomerism arises because rotation does not occur about a C=C double bond:

$$\underset{H}{\overset{Cl}{\diagdown}}C=C\underset{H}{\overset{Cl}{\diagup}}$$

$$\underset{H}{\overset{Cl}{\diagdown}}C=C\underset{Cl}{\overset{H}{\diagup}}$$

cis-1,2-dichloroethene   **trans**-1,2-dichloroethene

CLOSE-PACKING

Close-packing refers to the packing of identical spheres into an arrangement in which they occupy the minimum space. There are two kinds of close-packing. In each type any one sphere will be in contact with twelve other spheres. In **hexagonal close-packing**, the spheres in the third layer are directly above and repeat the arrangement of spheres in the first layer, giving the ABABAB pattern. In **cubic close-packing**, the spheres in the third layer are above the triangular holes between the spheres in the first layer, giving the ABCABC pattern. Most metals crystallise into either a hexagonal or a cubic close-packed structure. In the case of a cubic close-packed crystal, the metal atoms form a face-centred **unit cell** – see Fig. C.22.

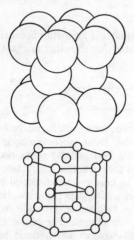

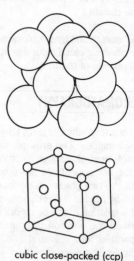

hexagonal close-packed (hcp)   cubic close-packed (ccp)

Fig. C.22 Two types of close-packing

COLLIGATIVE PROPERTY

A colligative property of a solution is a property which depends upon the **concentration** but not the nature of the solute particles.

At one time, the **elevation of the boiling point**, the **depression of the freezing point**, the **lowing of the vapour pressure** and the **osmotic pressure** of a solution were the four colligative properties of importance as a means of determining **relative molecular masses** of involatile compounds. For example, 1 dm^3 of water will freeze at $-0.19°C$ (instead of $0°C$) if 0.1 mol of a non-electrolyte is dissolved in it and at $-0.38°C$ if 0.2 mol of non-electrolyte is dissolved in it. By measuring the depression of the freezing point caused by dissolving a measured mass of solute in a measured mass of water, the relative molecular mass of the solute could be calculated.

Nowadays, more accurate and convenient methods are usually used for measuring relative molecular masses. However, osmotic pressure measurements may still be employed to determine average molar masses of **polymers**.

COLLISION THEORY

The collision theory is a theory developed to account for the effect of **concentration** and temperature upon the rates of gas-phase reactions in terms of the number of effective collisions occuring between the molecules of reactants. Expressed in simple terms, the theory assumes that a reaction will occur between two molecules only if they collide with a certain minimum amount of energy known as the **activation energy**. The **rate constant** for a reaction is related to the frequency factor (a measure of the collision rate) and the activation energy by the **Arrhenius equation**. Support for the collision theory was provided when the experimentally measured value of the rate constant for the decomposition of hydrogen iodide, for example, was found to match closely the theoretically calculated value.

◀ Order of reaction, Rate of reaction ▶

COLORIMETER

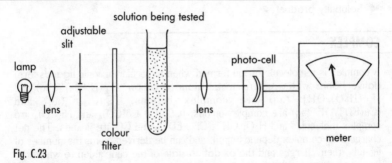

Fig. C.23

A colorimeter (Fig. C.23) is an instrument for comparing and measuring the intensity of electromagnetic radiation in the visible region of the spectrum. The colorimeter can be calibrated to relate the **concentration** of the coloured solute to the percentage of the light absorbed by the solution and the intensity of the light transmitted to the photocell and recorded on the meter.

COMMON-ION EFFECT

The common-ion effect is a term referring to the influence of the **concentration** of an aqueous ion upon an **equilibrium** involving that ion and a weak electrolyte or a sparingly soluble electrolyte.

The pH of $0.1 \text{ mol dm}^{-3} \text{ NH}_3(\text{aq})$ is approximately 11. If enough ammonium chloride is dissolved in the ammonia solution to give $0.1 \text{ mol dm}^{-3} \text{ NH}_4^+(\text{aq})$, the pH falls to 9. The addition of the common ion, $\text{NH}_4^+(\text{aq})$, alters the composition of the equilibrium:

$$NH_3(aq) + H_2O(l) \rightleftharpoons NH_4^+(aq) + OH^-(aq)$$

The increase in the concentration of the $\text{NH}_4^+(\text{aq})$ must be accompanied by a decrease in the concentration of the $\text{OH}^-(\text{aq})$ and an increase in the $\text{NH}_3(\text{aq})$ concentration to satisfy the law of chemical equilibrium for this **reversible reaction**; that is:

$$\frac{[NH_4^+(aq)][OH^-(aq)]}{[NH_3(aq)]} = K_b = 1 \times 10^{-5} \text{ mol dm}^{-3}$$

The solubility of silver chloride is approximately $10^{-5} \text{ mol dm}^{-3}$ AgCl in water and $10^{-9} \text{ mol dm}^{-3}$ AgCl in 0.1 mol dm^{-3} NaCl(aq). When silver chloride dissolves in water, an equilibrium is established between solid, undissolved silver chloride and its ions in solution:

$$AgCl(s) \rightleftharpoons Ag^+(aq) + Cl^-(aq)$$

In 0.1 mol dm^{-3} NaCl(aq), the concentration of the $\text{Cl}^-(\text{aq})$ is 0.1 mol dm^{-3}, so the concentration of the $\text{Ag}^+(\text{aq})$ must be $10^{-9} \text{ mol dm}^{-3}$ to satisfy the law of chemical equilibrium; that is:

$$[Ag^+(aq)][Cl^-(aq)] = K_{sp} = 1 \times 10^{-10} \text{ mol}^2 \text{ dm}^{-6}$$

◀ Solubility product ▶

COMPLEX

A complex is a molecule or ion formed when molecular or ionic **ligands** form **dative bonds** with a central metal **atom** or **cation**. $[Fe(H_2O)_6]^{3+}(\text{aq})$ and $[Fe(H_2O)_5OH]^{2+}(\text{aq})$ are complex cations, $[CuCl_4]^{2-}(\text{aq})$ and $[Cu(EDTA)]^{2-}(\text{aq})$ are complex anions, $(CO)_4Co–Co(CO)_4$ and $Ni(CO)_4$ are complex molecules and H_2O, OH^-, Cl^-, **EDTA** and CO are ligands. The net charge on a complex depends upon, and can be derived from, the number of ligands, their charges and the **oxidation state** of the central ion to which the ligands are bonded. Equally, the **oxidation number** of the central atom or ion can be derived from the formula and charge of the complex – see Fig. C.24.

---

1   What is the charge on the tetraamminedichlorochromium(III) complex ion?

the formula has two Cl^- atoms and four NH_3 molecules so the ligands contribute 2 negative charges

the central chromium cation has an oxidation number of $+3$ so it contributes 3 positive charges

hence the net charge on the complex ion is $1+$ and it is a cation:

$[CrCl_2(NH_3)_4]^+$

2   What is the oxidation number of copper in the $[Cu(CH)_4]^{3-}$ complex ion?

the complex ion is an anion with a net charge of $3-$

the formula has four CN^- anions as ligands that contribute 4 negative charges

hence the central copper ion must contribute 1 positive charge so its oxidation number is $+1$

$[Cu(CN)_4]^{3-}$ would be called the tetracyanocuprate(I) anion

---

Fig. C.24 Two simple problems on complex ions

Different ligands have different effects upon the stability of the oxidation state of a **transition metal**. Iodide ions are oxidised by hexaaquairon(III) ions but not by hexacyanoferrate(III) ions because the iron in the $+3$ oxidation state is more stable, and therefore a weaker **oxidising agent**, when complexed with CN^- rather than H_2O ligands:

$$
\begin{array}{ll}
 & E^\theta/\text{V} \\
[Fe(CN)_6]^{3-}(aq) + e^- \rightleftharpoons [Fe(CN)_6]^{2-}(aq) & +0.36 \\
\tfrac{1}{2}I_2(aq) + e^- \rightleftharpoons I^-(aq) & +0.54 \\
[Fe(H_2O)_6]^{3+}(aq) + e^- \rightleftharpoons [Fe(H_2O)_6]^{2+}(aq) & +0.77
\end{array}
$$

▶ NOMENCLATURE

The name of a complex starts with the names of the ligands in alphabetical order (ignoring numerical prefixes such as **di–**, **tetra–**, **hexa–**) and ends with the name and oxidation number of the central ion. The ending **–ate** identifies the complex as an anion. For example:

$[Cu(NH_3)_4(H_2O)_2]^{2+}(aq)$ tetraamminediaquacopper(II) ion

$[Fe(CN_6]^{3-}(aq)$ hexacyanoferrate(III) ion

$[Fe(CN)_6]^{4-}(aq)$ hexacyanoferrate(II) ion

$[Fe(SCN)(H_2O)_5]^{2+}(aq)$ pentaaquathiocyanatoiron(III) ion

Anionic ligands shown inside the [ ] of the formula are dative-covalently bound to the central ion and not free to act as anions.

▶ SHAPE

The number of atoms, from one or more ligands, bonded to the central atom or ion is called the **coordination number** and may be 2 or 4 (fairly common) or 6 (most common). The shape of a complex depends upon the coordination number and may be linear, tetrahedral, square planar or octahedral. Octahedral is the most common shape adopted even by the hexadentate EDTA complex ions (see Fig. C.25). Octahedral complexes can exhibit *cis–trans* (**geometrical**) isomerism and **optical isomerism** (see Fig. O.1).

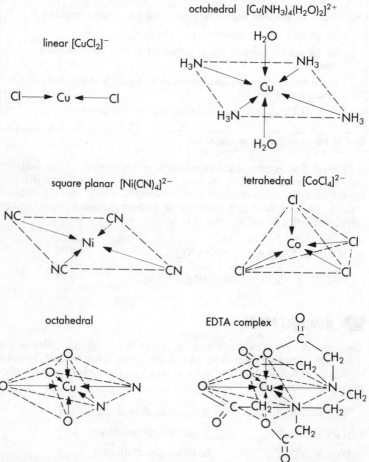

Fig. C.25 Shapes of complex ions

◤ COLOUR

Different ligands have different effects upon the colours of transition metal compounds and their **aqueous** solutions by changing the amount of energy, in the visible region of the spectrum, needed to excite the 3d-subshell electrons of the central transition metal atom or ion. Theoretically, this splitting of the 3d-subshell energy levels is treated by ligand-field theory, but this is outside the scope of A-level chemistry. Experimentally, these striking colour changes, when one ligand replaces another in a complex ion, may be observed in simple test-tube reactions employed in practical analytical chemistry: for instance, pale yellow aqueous iron(III) compounds become a vivid blood-red colour with aqueous thiocyanate ions:

$$[Fe(H_2O)_6]^{3+}(aq) + SCN^-(aq) \rightarrow [Fe(SCN)(H_2O)_5]^{2+}(aq) + H_2O(l)$$

pale yellow   blood red

◤ RELATIVE STABILITY

If concentrated hydrochloric acid is added to aqueous copper(II) sulphate, chloride ions replace water molecules as ligands to give a green solution containing yellow aqueous tetrachlorocuprate(II) anions and blue aqueous hexaaquacopper(II) cations:

$$4Cl^-(aq) + [Cu(H_2O)_6]^{2+}(aq) \rightleftharpoons [Cu(H_2O)_2Cl_4]^{2-}(aq) + 4H_2O(l)$$

If ammonia is added to this green solution, the colour turns deep blue; the tetraaminecopper(II) cation is more stable than the hexaaquacopper(II) cation and the diaquatetrachlorocuprate(II) anion:

$$4NH_3(aq) + [Cu(H_2O)_6]^{2+}(aq) \rightarrow [Cu(NH_3)_4(H_2O)_2]^{2+}(aq) + 4H_2O(l)$$

If aqueous EDTA (ethylenediaminetetraacetic acid) is added to the dark blue solution, the ammonia molecules are replaced by the hexadentate ligand to give a pale blue solution:

$$EDTA^{4-}(aq) + [Cu(NH_3)_4]^{2+}(aq) \rightarrow [Cu(EDTA)]^{2-}(aq) + 4NH_3(aq)$$

The increase in **entropy** accompanying the replacement of six monodentate ligands by one hexadentate ligand largely accounts for the high stability of EDTA complexes and for the use of EDTA in complexometric analysis.

◤ ACIDITY OF AQUA COMPLEX IONS

In aqua complex cations, the small central transition metal ion often has a sufficiently strongly polarising effect upon the datively bonded water molecules that these ligands lose protons and the complex acts as an acid – for example:

$$[Fe(H_2O)_6]^{3+}(aq) \rightleftharpoons [Fe(H_2O)_6OH]^{2+}(aq) + H^+(aq)$$

The **polarising power** of the central ion increases with increasing oxidation number, so aqua cations with a triple charge are much more acidic than those with a double charge. This explains why aqueous iron(III) salts but not aqueous iron(II) salts react with **carbonates** to produce carbon dioxide:

$$2H^+(aq) + CO_3^{2-}(aq) \rightarrow H_2O(l) + CO_2(g)$$

↑
protons from the hexaaquairon(III) cation

▶ HYDROLYSIS

When pale violet crystals of 'Iron(III)alum' (ammonium iron(III) sulphate-12-water) are dissolved in water, a yellow-brown solution is obtained; this solution gradually produces a brown precipitate:

$$[Fe(H_2O)_5OH]^{2+}(aq) \rightleftharpoons [Fe(H_2O)_4(OH)_2]^+(aq) + H^+(aq)$$

$$[Fe(H_2O)_4(OH)_2]^+(aq) \rightleftharpoons [Fe(H_2O)_3(OH)_3](s) + H^+(aq)$$

Acid is added to aqueous iron salts to minimise this **hydrolysis**. This is an application of the law of **chemical equilibrium** and the **common ion effect** because increasing the **concentration** of one product (the $H^+(aq)$ ion) must cause the concentration of the other products to decrease and that of the reactants to increase. Alkali is added to aqueous metal salts to maximise this hydrolysis and produce coloured precipitates that identify the cation – for example:

$$[Fe(H_2O)_6]^{2+}(aq) + 2OH^-(aq) \rightarrow [Fe(H_2O)_4(OH)_2](s) \qquad \text{dark green precipitate}$$

Even though these precipitates are more complicated than that shown by a simple formula like $Ni(OH)_2$, such complexities are beyond the requirements of A-level, and the equations are usually simplified as follows:

$$Cu^{2+}(aq) + 2OH^-(aq) \rightarrow Cu(OH)_2(s) \qquad \text{light blue precipitate}$$

CONCENTRATED SOLUTION

A concentrated solution is a solution which contains a large amount of solute in a small volume of the solution. Concentrated means the opposite of **dilute**. There is no clear dividing line between a concentrated solution and a dilute solution. For example, concentrated sulphuric acid contains about 2 per cent water, concentrated hydrochloric acid contains about 62 per cent water, and aqueous acids referred to as dilute acids may have concentrations up to 1 mol dm^{-3}. The term 'concentrated' must not be confused with the term 'strong'.

CONCENTRATION

Concentration is the amount of solute substance per unit volume of solution; it is measured in mol dm^{-3}. The term usually applies to a solid or liquid solute dissolved in a liquid solvent to give a liquid solution. The solute should be specified by its formula and name but not by its name alone. If 9.8 g of sulphuric acid (= 0.1 mol H_2SO_4) is dissolved in water to give 1 dm^3 of solution, the concentration of the solution may be written as $[H_2SO_4] = 0.100$ mol dm^{-3}, or as $c(H_2SO_4) = 0.100$ mol dm^{-3}. Although it is less acceptable, this can also be described as a 0.100 M H_2SO_4(aq) solution or a solution whose molarity with respect to H_2SO_4 is 0.100. However, the term **molarity** is best avoided and must not be confused with **molality**.

Mass (as opposed to *amount of substance*) concentration refers to the mass of solute per unit volume of solution. It is expressed in g dm^{-3} or kg dm^{-3} and has the symbol ρ. Mass concentration is sometimes expressed as per cent, W/V as, for example, with the description of commercial 'concentrated' hydrochloric acid as 38 per cent W/V solution because it contains 38 g of HCl in 100 cm^3 of solution. Solubility, which is a measure of mass concentration, is often expressed in g of solute per 100 g of solvent.

CONDENSATION POLYMER

A condensation polymer is a **polymer** formally derived from **monomers** joined by a **condensation reaction** involving the elimination of water or some other simple molecule. **Polyamides**, **polyesters** and **polysaccharides** are three important broad classes of condensation polymers. The polyamides include the synthetic **nylons** and the naturally occurring **proteins**. The polysaccharides include starch and cellulose.

CONDENSATION REACTION

A condensation reaction is the combination of two molecules to form a larger molecule and expel a smaller molecule. It is usually regarded as an **addition reaction** followed by an **elimination reaction**.

Acyl chlorides react with **alcohols**, **phenols**, ammonia and **amines** to form **esters** and **amides**; see Fig. A.1, page 1.

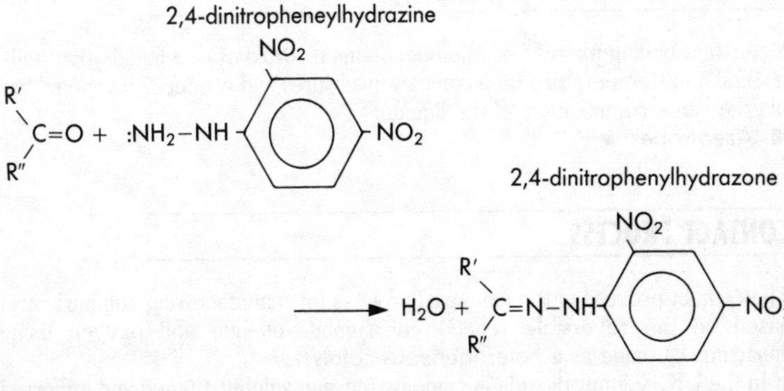

Fig. C.26 Formation of 2,4-dinitrophenylhydrazone derivations

Aldehydes and ketones react with derivatives of ammonia such as hydroxylamine, NH_2OH, and 2,4-dinitrophenylhydrazine, $(NO_2)_2C_6H_3NHNH_2$, to form well-defined condensation products – see Fig. C.26

CONJUGATE ACID

A conjugate acid is the **acid** formed when a **base** accepts a **proton**. For example, according to the Bronsted–Lowry theory, the ammonia molecule and the **hydroxide ion** can act as bases by accepting a proton. When they do so, they form their conjugate acids, the ammonium ion and the water molecule, respectively:

$$NH_3(aq) + H^+(aq) \rightarrow NH_4^+(aq) \qquad OH^-(aq) + H^+(aq) \rightarrow H_2O(l)$$

↑ base	↑ conjugate acid	↑ base	↑ conjugate acid

CONJUGATE BASE

A conjugate base is the **base** formed when an **acid** loses a **proton**. For example, according to the Bronsted–Lowry theory, the hydronium ion and the sulphuric acid molecule can act as acids by losing a proton. When they do so, they form their conjugate bases, the water molecule and the hydrogensulphate ion, respectively:

$$H_3O^+(aq) \rightarrow H^+(aq) + H_2O(l) \qquad H_2SO_4(aq) \rightarrow H^+(aq) + HSO_4^-(aq)$$

↑ acid	↑ conjugate base	↑ acid	↑ conjugate base

CONSTANT BOILING MIXTURE

A constant boiling mixture is a homogeneous mixture of two liquids that boils at a constant temperature (at a constant pressure) and produces a vapour that has the same composition as the liquid.
◀ Azeotropes ▶

CONTACT PROCESS

The contact process is the industrial process for manufacturing sulphuric acid based on the **reversible** reaction of sulphur dioxide and oxygen using vanadium(V) oxide as a **heterogeneous catalyst**.

In the UK, sulphur dioxide is made by burning sulphur (mined and imported in the liquid state) in dry air. The gas is filtered and cooled to about 690 K, mixed with twice the amount of oxygen prescribed by the following equation and passed through beds of solid vanadium(V) oxide catalyst pellets:

$$2SO_2(g) + O_2(g) \rightleftharpoons 2SO_3(g); \quad \triangle H^\theta = -98kJ \ mol^{-1}$$

Heat exchangers remove the heat evolved to maintain the catalyst at the optimum temperature of around 690 K. About 99 per cent of the sulphur dioxide is converted to the trioxide which is absorbed in sulphuric acid containing 2 per cent water:

$$SO_3(g) + H_2O(l) \rightarrow H_2SO_4(l)$$

COORDINATE BOND

A coordinate bond is a **covalent bond** in which one of the **atoms** has supplied both **electrons** being shared. Coordinate bonds are also called dative (and sometimes co-ionic) bonds. Once a coordinate bond is formed, the sharing of the electron pair would be indistinguishable from a covalent bond. Compounds containing **complex** ions and molecules are sometimes known as coordination compounds because the **ligands** are attached to the central atom or ion by coordinate bonds.

◀ Complex ▶

COORDINATION NUMBER

The coordination number of an atom or ion is the number of nearest equidistant atoms, molecules, ions or groups in a crystal structure or in a **complex**.

In the **hexaganol** or **cubic close-packed crystal structures** formed by most metallic elements, the coordination number is 12. In the **body-centred structure** typical of the **alkali metals**, the coordination number is 8. In crystals of ionic compounds, the coordination numbers of the ions may be 8, 6 or 4. They are determined by the relative sizes of the **anion** and **cation** and by their charges. For example, in the interpenetrating simple cubic structure of **caesium chloride** (CsCl), each ion has a coordination number of 8 and in the interpenetrating **face-centred cubic** (rock-salt) structure of sodium chloride (NaCl), each ion has a coordination number of 6.

The ratio of the coordination numbers of the **cation** and **anion** gives the ratio of the amounts of ions in the compound and hence the **empirical formula**. In the fluorite structure the calcium ion is in the centre of a cube formed by eight fluoride ions at its corners and the fluoride ion is at the centre of a tetrahedron formed by four calcium ions at its corners. The ratio of the coordination numbers is 8 (F^- ions) to 4 (Ca^{2+} ions) and the empirical formula is CaF_2.

In simple and giant molecular covalent crystal strucures, the coordination number is determined by the number of bonds formed between the **atoms**. For example, carbon is capable of forming four **covalent bonds**. In the diamond structure each carbon atom is bonded to four other carbon atoms. In the iodine crystal, the I_2 molecules are arranged in a fairly close-packed structure but the nuclei of the two atoms in one molecule are closer together than the nuclei of two atoms in two separate molecules.

In a **complex** ion or molecule, the coordination number of the central atom or ion is frequently 6, and the six **dative** (coordinate) **bonds** are directed towards the corners of an octahedron as, for example, in the hexaaquacobalt(II) cation. The coordination number may also be 4 or 2 – as, for example, in the tetrahedral $CoCl_4^{2-}$, square planar $CoCl_4^{2-}$ and linear $CuCl_2^-$ complex anions.

COUPLING REACTION

A coupling reaction is an **electrophilic substitution reaction** between a diazonium ion (the **electrophile**) and a coupling agent (the **nucleophile**) to form an **azo** dye. The diazonium ions are produced by **diazotisation** of primary amino groups attached directly to an **aromatic** ring using nitrous acid in ice-cold conditions. Coupling agents include substituted **phenols**, naphthols and aromatic **amines**.

◀ Azo dyes ▶

COVALENT BONDING

A covalent bond forms between two **atoms** of non-metallic elements when the nuclear attraction for the shared **electrons** outweighs the repulsions between the **nuclei** and between the electrons of two atoms.

Single, double and triple covalent bonds between two atoms involve the sharing of two, four and six electrons respectively. According to the **octet rule**, the number of pairs of electrons being shared, and therefore the number of covalent bonds formed, between two atoms will usually be 8– the non-metal's group number in the **periodic table**. For example, nitrogen is in Group V so in N_2 the two atoms are joined by a triple bond.

A covalent bond can be represented in a diagram using 'dots and crosses' to keep count and show the origin of the **valence electrons** in each atom. If one atom supplies both electrons for sharing, the covalent bond is called a **dative** or coordinate **bond**. Once the bond is formed, the electrons being shared are not distinguishable by their origin. A shared or bonded pair of electrons is more usually represented by a single line drawn between the symbols of the two covalently bonded atoms. Lone or non-bonded pairs of electrons are, when necessary, represented by two dots (see Fig. D.2).

The directional character of a covalent bond and the repulsions between (bonded and non-bonded) pairs of valence electrons can account for the shapes of simple molecules. Covalent bonds have definite **bond energies** whose values depend upon the nature of the atoms, X and Y, sharing the electrons, the number of electrons being shared and the other atoms or groups attached to X and Y. If X and Y have different **electronegativities**, the covalent bond will be polar because the atoms will share the electrons unequally.

◀ Dative bonding, Double bond, Single bond, Triple bond ▶

COVALENT RADIUS

In a simple **diatomic molecule** the covalent radius is half the distance between the two nuclei.
◀ Atomic radius ▶

CRACKING

Cracking is the breaking down of large **molecules** into smaller ones by the action of heat, with or without the aid of a **catalyst**. The process is of major importance to the petrochemical industry in the production of branched-chain **alkanes** and **aromatic hydrocarbons** for petrol with a high 'octane rating' (good 'anti- knock' characteristics) and the supply of a feedstock of **alkenes** for the chemical industry. Catalytic cracking and thermal cracking are complicated molecular decompositions, rearrangements and recombinations involving **free radical chain mechanisms**.

Alkenes are produced by cracking (and **dehydrogenation**) of ethane and propane (from natural gas) and other alkanes (from light petroleum distillates in the naphtha range) – for instance:

$$CH_3-CH_3 \xrightarrow{\quad -H_2 \quad} CH_2=CH_2$$
$$C_3H_8 \xrightarrow{\quad\quad} CH_2=CH_2 + CH_4$$

CRYSTAL

A crystal is a solid substance with a regular shape having flat surfaces, sharp edges and one, two or three definite angles between pairs of faces.

Crystals have a sharp melting point, a refractive index and constant interfacial angles which may be used to identify the substance. The formation of crystals by the process of **crystallisation** from a liquid or solution provides an important method of purification. Crystals give distinctive **X-ray diffraction** patterns from which the dimensions and composition of the **unit cell** of the **crystal lattice** may be determined.

On the basis of their unit cell, crystalline substances can be grouped into seven major crystal systems: cubic, tetragonal, orthorhombic, rhombohedral, hexagonal, monoclinic and triclinic. Most metals form **cubic** or **hexagonal close-packed** crystals. The **alkali metals** form **body-centred** cubic crystals. The alkali metal halides form cubic crystals that are **isotropic**. Calcite, a rhombohedral crystal of calcium carbonate, is **anisotropic**. Crystals of non-metallic elements and of compounds are usually non-conducting, brittle and easily cleaved.

CRYSTAL LATTICE

A crystal lattice is the regular arrangement of **atoms, ions** or molecules in a crystalline substance seen as a pattern produced by the repeated translation of the **unit cell**.

CRYSTALLISATION

Crystallisation is the formation of crystals from a gas, liquid or **supersaturated** solution.

CRYSTAL STRUCTURE

The crystal structure of a substance is the regular spatial arrangement of its **atoms, ions** or **molecules** in a **crystal lattice**.

CUBIC CLOSE-PACKING

Cubic close-packing is the packing of identical metal atoms to occupy the minimum amount of space and to produce a **coordination number** of 12 and a **face-centred cubic unit cell**.
◄ Close-packing ►

CYCLOALKANES

Cycloalkanes are a **homologous series** of **cyclic saturated hydrocarbons** having the general formula C_nH_{2n}. Although they have the same general formula as the **alkenes**, they do not contain a $C=C$ **double bond**. The carbon atoms in the ring are joined by **single covalent bonds**. Consequently, cycloalkanes do NOT undergo **electrophilic addition reactions** with **halogens**. Cyclopropane is the simplest homologue:

$$CH_2-CH_2$$
$$\diagdown \diagup$$
$$CH_2$$

DALTON'S LAW OF PARTIAL PRESSURES

In a mixture of gases, each gas exerts its pressure independently as if the other gases were not present, so that the total pressure (P) of the mixture equals the sum of these **partial pressures** p_A, p_B, and so on:

$$P = p_A + p_B$$

Dalton's law of partial pressures is strictly true only for **ideal gas** mixtures. Consequently the law is likely to be disobeyed at high pressures and low temperatures. The law does not hold for mixtures of gases that react together.

◀ Raoult's law ▶

DANIELL CELL

The Daniell cell (Fig. D.1) is an **electrochemical cell** consisting of a **half-cell** of zinc metal dipping into **aqueous** zinc sulphate in electrolytic contact with a

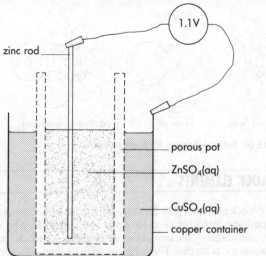

Fig. D.1

half-cell of copper metal dipping into aqueous copper(II) sulphate. This electrochemical cell was invented in 1836 by a British chemist, John Daniell. It often consists of an unglazed porcelain porous pot filled with saturated aqueous zinc sulphate and standing in a copper container filled with saturated aqueous copper(II) sulphate. Under standard conditions with the liquid–liquid junction potential eliminated, the e.m.f. of the Daniell cell is 1.10 V –

$$Zn(s)\left|Zn^{2+}(aq)\right|\left|Cu^{2+}(aq)\right|Cu(s) \quad E^{\theta} = +1.10 \text{ V}$$

When the cell is short-circuited and delivering current, the cell reaction is the **displacement** of aqueous copper(II) cations by zinc:

$$Zn(s) + Cu^{2+}(aq) \rightarrow Zn^{2+}(aq) + Cu(s)$$

DATIVE BONDING

A dative (or coordinate) bond is a **covalent bond** in which one of the **atoms** has supplied both **electrons** being shared.

Although 'dots and crosses' may be used in diagrams (see Fig. D.2) to represent the outer **valence electrons** (to show their origin and count them), once the bond is formed the electrons are not distinguishable by their origin. The atom donating the **lone pair** of electrons may be called a **Lewis base**. The atom accepting the lone pair is called a **Lewis acid**.

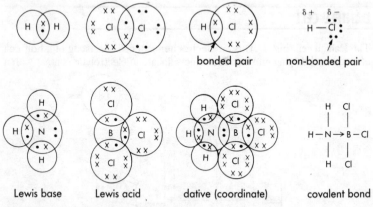

Fig. D.2 Dot and cross diagrams for simple molecules

d-BLOCK ELEMENTS

The d-block elements (Fig. D.3) are the metals in the horizontal rows crossing the **periodic table** between Group II and Group III. The d-block elements have remarkably similar physical properties, which are regarded as characteristic of metals. The hardness, high melting points and high enthalpies

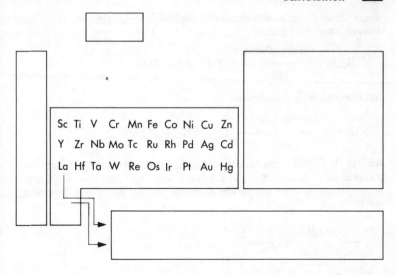

Sc	Ti	V	Cr	Mn	Fe	Co	Ni	Cu	Zn
Y	Zr	Nb	Mo	Tc	Ru	Rh	Pd	Ag	Cd
La	Hf	Ta	W	Re	Os	Ir	Pt	Au	Hg

Fig. D.3

of melting indicate very strong **metallic bonding** involving **delocalisation** of the inner d-subshell and outer s-subshell electrons. They form alloys and can act as **heterogeneous catalysts**.

The formation of coloured compounds and **complex ions** in various **oxidation states** is an important general feature of the chemistry of the d-block elements. For A-level chemistry, the emphasis is on the first row of d-block elements from scandium to zinc and the first series of **transition metals** from titanium to copper.

◀ Transition metals ▶

DEGREE OF DISSOCIATION

The degree of dissociation, α, is the fraction of the total amount of **molecules** or **ions** that have split into smaller molecules or ions in a **reversible** dissociation **reaction**.

◀ Dissociation constant ▶

DEHYDRATION

Dehydration is the removal of water or the elements of water from a substance. Blue hydrated copper(II) sulphate-5-water crystals are dehydrated to white anhydrous copper(II) sulphate powder when gently heated:

$$CuSO_4 \ 5H_2O(s) \rightarrow CuSO_4(s) + 5H_2O(l); \ \triangle H^{\theta} = + \ 1508kJ \ mol^{-1}$$

When heated with concentrated sulphuric acid, ethanol undergoes *intermolecular* dehydration:

$$2C_2H_5OH \xrightarrow[140°C]{\text{excess ethanol}} C_2H_5OC_2H_5 + H_2O$$

and *intramolecular* dehydration:

$$2C_2H_5OH \xrightarrow[160°C]{\text{excess } H_2SO_4(l)} CH_2{=}CH_2 + H_2O$$

Amides ($R–CONH_2$) can be dehydrated to **nitriles** ($R–CN$) by heating with phosphorus(V) oxide. Benzene-1,2-dicarboxylic acid (phthalic acid) dehydrates to the internal **anhydride** on gentle heating just above its melting point:

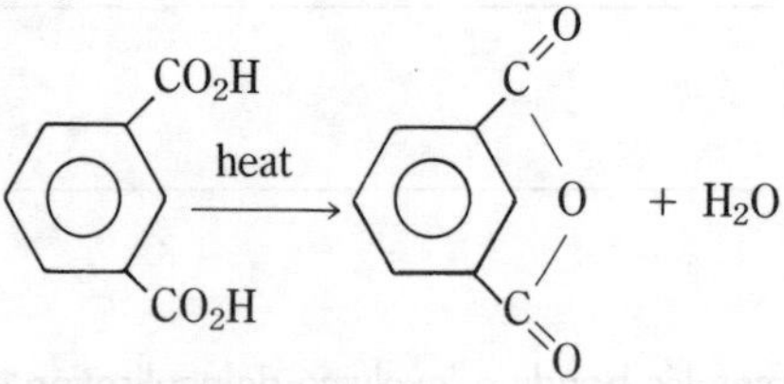

DEHYDROGENATION

Dehydrogenation is the removal of gaseous hydrogen from a gaseous compound using a **heterogeneous catalyst**.

In North America, butadiene required for the industrial production of synthetic rubber is obtained by dehydrogenation of butane:

$$CH_3CH_2CH_2CH_3 \xrightarrow[Al_2O_3/Cr_2O_3 \text{ catalyst}]{600°C \text{ reduced press}} CH_2{=}CHCH{=}CH_2 + 2H_2$$

In the UK, one method of manufacturing propanone is the dehydrogenation of propan-2-ol using a heterogeneous catalyst of either zinc oxide on pumice at 380°C or copper metal at 550°C:

$$CH_3CH(OH)CH_3 \rightarrow CH_3COCH_3 + H_2$$

DEHYDROHALOGENATION

Dehydrohalogenation is the removal of hydrogen halide from a molecule by an **elimination reaction**. The overall elimination reaction may be represented by the following general equation:

$$\underset{\displaystyle \overset{|}{\underset{|}{-C}}-\overset{|}{\underset{|}{C}}-}{\overset{H \quad X}{}} + OH^- \rightarrow {>}C{=}C{<} + X^- + H_2O$$

Samples of **alkenes** such as 2-methylpropene can be prepared in the laboratory by heating the appropriate **halogenoalkane** (2-chloro-2-methylpropane) with a **concentrated** solution of potassium hydroxide in ethanol – see Fig. D.4.

gaseous 2-methylpropene collected over water

glass wool to support mixture of ethanolic potassium hydroxide and 2-chloro-2-methylpropane

heat

$$CH_3-\overset{\overset{\displaystyle Cl}{|}}{\underset{\underset{\displaystyle CH_3}{|}}{C}}-CH_3 + KOH \xrightarrow{\text{heat in} \atop \text{ethanol}} \overset{\displaystyle CH_3}{\underset{\displaystyle CH_3}{>}}C=CH_2 + KCl + H_2O$$

Fig. D.4 Laboratory preparation of 2-methylpropene

DELOCALISATION

Delocalisation is a term which refers to **valence electrons** being distributed over three or more atoms rather than being restricted to bonds between two atoms.

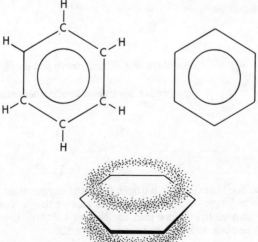

Fig. D.5 Delocalised π-electrons in benzene

DELOCALISATION

Benzene is a good example of a molecule which has delocalised electrons in its structure. Each carbon atom in the ring is joined to its neighbours and to a hydrogen atom by σ-bonds. This accounts for three out of the four valence electrons of each carbon atom. The remaining six valence electrons (one from each carbon atom) are distributed over all six carbon atoms to form delocalised σ-bonding above and below the plane of the ring – see Fig. D.5. If more than one way can be found of allocating the valence electrons to the bonding of atoms in a molecule or ion without contravening the rules of bonding, some of the electrons will probably be delocalised – see Fig. D.6.

Write down the arrangement of the symbols and the number of valence electrons of each atom then alter the numbers as you draw in the bonds between the symbols to give H-atoms 2 and other atoms 8 electrons to complete their outer shells:

full shell   full shell

$$
\begin{array}{cccc}
1 & 6 & 5 & 6 \\
H & O & N & O \\
 & & O & \\
 & & 6 &
\end{array}
$$

$$
\begin{array}{cccc}
 & \downarrow & & \\
2 & 7 & 5 & 6 \\
\rightarrow H{-}O & N & O \\
 & O & \\
 & 6 &
\end{array}
\qquad
\begin{array}{cccc}
 & & \downarrow & \\
2 & 8 & 6 & 6 \\
\rightarrow H{-}O{-}N & O \\
 & O & \\
 & 6 &
\end{array}
$$

full shell so lone pair forms dative bond

$$
\begin{array}{cc}
\downarrow & \\
8 & 6 \\
H{-}O{-}N: & O \\
\|| & \\
O & \\
8 & \\
\uparrow & \\
\text{full shell}
\end{array}
\qquad
\begin{array}{cc}
8 & 8 \\
\rightarrow H{-}O{-}N\rightarrow O \\
\|| \\
O \\
8
\end{array}
\qquad
\begin{array}{cc}
8 & 8 \\
H{-}O{-}N{=}O \\
\downarrow \\
O \\
8
\end{array}
$$

two ways of allocating dative and double bond

$\Rightarrow$ predict delocalised bonding between N and O

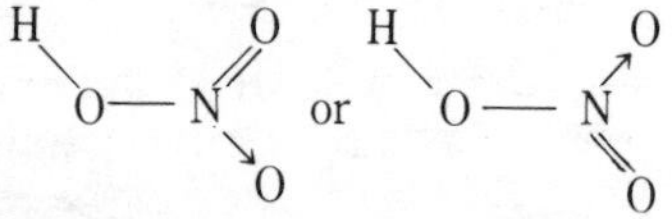

You predict the bond between the nitrogen atom and oxygen atom to be identical and to be longer than a N=O bond but shorter than a N–O bond. And since the nitrogen has no lone pair, the N-atom and three O-atoms will all be in the same plane with an ONO bond angle of 120°.

Fig. D.6 Predicting delocalised bonding in nitric acid

When the nitric acid molecule ionises to NO_3^-, all three NO bonds become equal and the negative charge is delocalised across all three oxygen atoms. In simple ions or molecules, the delocalised electrons are confined within the molecule or ion. In a metal the electrons are delocalised over all the atoms throughout the structure.

◀ Metallic bonding ▶

DENSITY

Density (ρ) is the concentration of matter as measured by the mass per unit volume. Density of solids and liquids is conveniently measured in the units g cm^{-3} and that of gases in g dm^{-3}. The coherent SI units are kg m^{-3}.

DEPRESSION OF THE FREEZING POINT

The freezing point depression is the lowering of the temperature at which a liquid solidifies when a substance is dissolved in it. The fall ($\triangle t$) in the freezing point is a **colligative property** dependent upon the **concentration** (C_m) but not the nature of the solute particles according to the relationship $\triangle t = K_f C_m$. K_f (units K kg mol^{-1}) is the cryoscopic constant for the liquid acting as a solvent. Measurement of the depression of freezing point of a measured mass of solute has been used to determine the **molar mass** of the solute.

DETERGENTS

A detergent is a *surfactant* (surface-active agent) which lowers the surface tension of water and aids cleaning by emulsifying and solubilising grease and dirt.

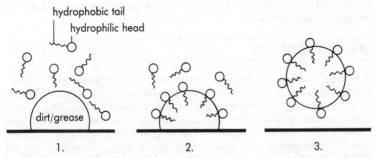

Fig. D.7 Essential action of a detergent

Detergents may be classed as soapy and soapless. **Soaps** are anionic detergents, being usually sodium salts of long-chain **fatty acids**, such as sodium octadecanoate (sodium stearate) $CH_3(CH_2)_{16}CO_2^-Na^+$. Soapless detergents may be anionic (such as sodium dodecylbenzenesulphonate

$CH_3(CH_2)_{11}C_6H_4 SO_3^-Na^+$), cationic (such as trimethylpentadecylammonium bromide $[(CH_3)_3N(CH_2)_{14}CH_3]^+Br^-$) or nonionic (such as poly(epoxyethanes) $R-CH_2-O-(CH_2CH_2O)_n-CH_2CH_2OH$). In any detergent the hydrocarbon 'tail' of the **ion** or **molecule** is **hydrophobic** (water-hating) or *lipophilic* (fat-loving) and the ionic or polar 'head' is **hydrophilic** or *lipophobic*. The essential action of a detergent is illustrated in Fig. D.7.
◀ Soaps ▶

DEXTROROTATORY

A dextrorotatory substance is an optically active substance which rotates the plane of **polarised light** to the right in a clockwise direction as viewed from a position facing the plane-polarised light source.
◀ Optical acitivity ▶

DIAGONAL RELATIONSHIP

A diagonal relationship is a chemical similarity between a second period element and a third period element in the next **group** of the **periodic table**.

From left to right across the **period** (Li to F) metallic character decreases and non-metallic character increases, while from top to bottom down the groups (Li to Cs, Be to Ba, and so on) metallic character increases and non-metallic character decreases. These trends result in pairs of elements in neighbouring groups having similar properties. Li and Mg, Be and Al, B and Si, and Cl and O are the most frequently quoted pairs of elements showing diagonal relationships. For example, lithium and magnesium both burn in air to form the normal **oxide** only. On heating in a bunsen flame, their **carbonates** decompose into the oxide and carbon dioxide. Their **chlorides** are soluble in ethanol and slowly hydrolysed by water. These properties are not shown by the other **alkali metals**.

DIATOMIC MOLECULE

A diatomic molecule is a molecule composed of two covalently bonded atoms of either the same or different elements. Gaseous elements, such as hydrogen (H_2), oxygen (O_2) and nitrogen (N_2), the **halogens** (F_2, Cl_2, Br_2, I_2) and hydrogen halides (HCl, HBr, HI), and the monoxides of carbon (CO) and nitrogen (NO), are all diatomic. The **noble gases** (He, Ne, Ar, Kr, Xe) are monatomic (atomicity = 1), while carbon dioxide (CO_2), ammonia (NH_3) and (CH_4) methane have an atomicity of 3, 4 and 5 respectively.

DIAZONIUM ION

A diazonium ion is a **cation** with the general formula $R-N_2^+$ formed by the **diazotisation** of a primary amino group by nitrous acid.

π-electrons of the bond between the nitrogen atoms become part of the delocalised π-electron system of the benzene ring and help to stabilise the diazonium ion

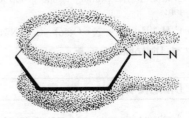

Fig. D.8 Benzene diazonium ion stabilised by delocalisation

Aliphatic diazonium ions are extremely unstable, decomposing readily into N_2 gas and a variety of other products. In ice-cold conditions, aromatic diazonium ions are stabilised by **delocalisation**, the π-electrons of the nitrogen–nitrogen bond participating in the delocalised π-electron system of the benzene ring (see Fig. D.8). On heating, the aqueous diazonium ion decomposes into nitrogen gas following **first-order reaction** kinetics:

$$R-N_2^+ \rightarrow R^+ + N_2(g)$$

and the **carbocation**, R^+, combines with or attacks various **Lewis bases** to form various other products, including **phenol**, which can be detected by its smell. If the solution is warmed with aqueous potassium iodide, iodobenzene is formed:

$$C_6H_5-N_2^+ + I^- \rightarrow C_6H_5-I + N_2(g)$$

If the solution is warmed with aqueous potassium cyanide and copper(I) cyanide as **catalyst**, benzonitrile is formed:

$$C_6H_5-N_2^+ + CN^- \xrightarrow[\text{catalyst}]{CuCN} C_6H_5-CN + N_2(g)$$

These reactions are important in providing one of the few ways of attaching I or CN to a benzene ring. The most important use of diazonium ions is the production of **azo-dyes**.

DIAZOTISATION

Diazotisation is the reaction of ice-cold aqueous nitrous acid with an **aromatic** primary amine to form an aqueous diazonium compound. The weak nitrous acid, formed *in situ* on mixing hydrochloric acid and aqueous sodium nitrite, undergoes protonation followed by **heterolytic fission** to form the electrophilic nitrosonium ion:

$$H_3\overset{\bullet}{O}{}^+(aq) + HNO_2(aq) \rightarrow H_2O(l) + H_2O\overset{+}{-}N=O(aq) \rightarrow 2H_2O(l) + \overset{+}{N}=O$$

Reaction between the electrophilic nitrosonium ion and the nucleophilic N-atom in the primary amino group leads to the formation of the diazonium ion:

$$C_6H_5NH_2(aq) + \overset{+}{N}{=}O(aq) \rightarrow C_6H_5N_2^+(aq) + H_2O(l)$$

◄ Diazonium ion ►

DIBASIC ACID

A dibasic (diprotic) acid means an **acid** such as sulphuric acid (H_2SO_4) with two moles of donatable **protons** per **mole** of the acid. One mole of a dibasic acid will react with one and two moles of a (monacidic) base such as sodium hydroxide to form the acid and the normal salt respectively:

$$H_2SO_4(aq) + NaOH(aq) \rightarrow H_2O(l) + NaHSO_4(aq) \qquad \text{sodium hydrogensulphate}$$

$$H_2SO_4(aq) + 2NaOH(aq) \rightarrow 2H_2O(l) + Na_2SO_4(aq) \qquad \text{sodium sulphate}$$

The term is not required by the **Bronsted–Lowry theory**, which defines all acids as monobasic. For example, H_2SO_4 is an acid which donates a proton to form its **conjugate base** the hydrogensulphate ion, HSO_4^-:

$$H_2SO_4 \rightarrow H^+ + HSO_4^-$$

and the hydrogensulphate ion is an acid which donates a proton to form its conjugate base, the sulphate ion, SO_4^{2-}:

$$HSO_4^- \rightarrow H^+ + SO_4^{2-}$$

The hydrogensulphate ion is therefore seen to be amphoteric.

DIFFUSION

Diffusion is the process involved when substances form uniform mixtures as a result of the continual random movement of their particles.

Gases diffuse rapidly, **Graham's law** of diffusion and measurement of the rate of gaseous diffusion (or effusion) provided a way of determining the **relative molecular masses** of gases. The difference in the rates of diffusion of $^{235}UF_6$ and $^{238}UF_6$ was the basis of a method of separating the uranium isotopes. Diffusion in liquids is slow and in solids it is almost negligible.

◄ Graham's law ►

DILUTE

A dilute solution is one which contains a small amount of solute in a large volume of the solution. Dilute means the opposite of **concentrated**. There is no clear dividing line between a dilute and a concentrated solution. For example, aqueous solutions referred to as 'dilute' may have concentrations up

to 1 mol dm^{-3}. A dilute solution is sometimes defined as a solution in which the mole fraction of the solvent closely approaches 1. The term 'dilute' must not be confused with the term 'weak' when applied to acids and bases. A solution containing 0.0001 mol dm^{-3} HCl(aq) would be described as a dilute solution of a **strong acid**. A solution containing 5 mol dm^{-3} CH$_3$CO$_2$H(aq) would be a concentrated solution of a **weak acid**.
◀ Concentrated ▶

DIMER

A dimer consists of two identical **molecules** bonded together. The molecules may be held together by weak forces such as **hydrogen bonds** when, for example, molecules of **carboxylic acid** dimerise in a non-polar solvent or in the liquid or crystalline state. The molecules may also be held together by **covalent bonds** – as, for example, is the case with aluminium chloride in the vapour state:

$$CH_3C \overset{O \cdots H-O}{\underset{O-H \cdots O}{\diagup\diagdown}} CCH_3 \qquad \underset{Cl \ Cl \ Cl}{\overset{Cl \ Cl \ Cl}{Al \ Al}}$$

DIMERISATION

Dimerisation is the combining of two identical **molecules** to form a dimer.
◀ Dimer ▶

DIPOLE–DIPOLE ATTRACTIONS

Dipole–dipole forces are the attractions arising between the negative end of one permanently **polar molecule** and the positive end of another permanently polar molecule.

These dipole–dipole forces can be distinguished from two other kinds of intermolecular dipole–dipole attraction called *London* or *dispersion* forces: a) attraction between the permanent dipole of a polar molecule and a temporary dipole induced in a neighbouring non-polar molecule; b) attraction between instantaneously induced dipoles in non-polar molecules resulting from the movement of electrons in relation to the nuclei in atoms.

The term **van der Waals forces** sometimes refers to all three types of dipole–dipole attraction and sometimes applies only to the instantaneous attractions of induced dipoles between non-polar molecules.
◀ Van der Waals forces ▶

DIPOLE MOMENT

A dipole moment (p) is a measure of the polarity of a **molecule** containing polar covalent bonds between one or more pairs of atoms with different electronegativity index values.

In a diatomic molecule the more electronegative atom has a slight negative charge ($\delta-$) and the less electronegative atom has a slight positive charge. The dipole moment is the product of one of these slight charges and the distance between them. If $\delta+$ (or $\delta-$) is measured in coulombs and the distance is in metres, the units of p would be coulomb metres (C m). The usual unit is the debye (D) where $1\ D \simeq 3.34 \times 10^{-30}$ C m.

Dipole moments can give information about the structure and shape of a molecule. The dipole moment of carbon dioxide (0 D) suggests a linear structure for the CO_2 molecule in which the equally polar $\delta+C=O\delta-$ bonds act in exactly opposite directions – see Fig. D.9.

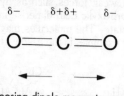

$$\delta- \qquad \delta+\delta+ \qquad \delta-$$

$$O = C = O$$

opposing dipole moments cancel

Fig. D.9 Linear non-polar molecule

The dipole moment of water (1.84 D) suggests that the water molecule cannot be linear, but must be bent so that both the equally polar $\delta+H-O\delta-$ bonds are pointing in similar directions – see Fig. D.10.

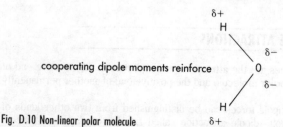

cooperating dipole moments reinforce

Fig. D.10 Non-linear polar molecule

DISPLACEMENT

A displacement reaction is the reaction of a reactive element with the compound of a less reactive element. For instance, zinc is more reactive and higher in the electrochemical series than copper, so zinc will displace copper from aqueous copper(II) sulphate in a reaction which is the basis of the **Daniell cell**:

$$Zn(s) + Cu^{2+}(aq) \rightarrow Zn^{2+}(aq) + Cu(s)$$

Sodium and magnesium are more reactive than titanium, so they are used to produce titanium metal by a displacement reaction with titanium(IV) chloride.

Aluminium is used in the 'thermit' reaction to displace iron from iron(III) oxide for the *in situ* welding of railway lines:

$$2Al(s) + Fe_2O_3(s) \rightarrow Al_2O_3(s) + 2Fe(s)$$

Chlorine is more reactive than bromine, so chlorine is used to displace bromine from aqueous bromide ions in the commercial extraction of bromine from sea water:

$$Cl_2(aq) + 2Br^-(aq) \rightarrow 2Cl^-(aq) + Br_2(aq)$$

An *acid displacement reaction* is the reaction of a strong acid with the conjugate base of a weaker acid. Mineral acids and carboxylic acids, but not phenol, will displace carbonic acid from carbonates and hydrogencarbonates to produce carbon dioxide. For instance:

$$H_2SO_4(aq) + Na_2CO_3(s) \rightarrow Na_2SO_4(aq) + H_2CO_3(aq) \rightarrow H_2O(l) + CO_2(g)$$

An *involatile acid* (or base) will displace a more volatile acid (or base) from its salt. Hydrogen chloride (or ammonia) can be displaced from sodium chloride (or ammonium sulphate) by phosphoric acid (or calcium hydroxide):

$$H_3PO_4(l) + NaCl(s) \xrightarrow{\text{heat}} NaH_2PO_4(s) + HCl(g)$$

$$Ca(OH)_2(s) + (NH_4)_2SO_4(s) \xrightarrow{\text{heat}} CaSO_4(s) + 2H_2O(l) + 2NH_3(g)$$

Organic substitution reactions are sometimes described as displacements.
◀ Alkali metals, Alkaline earth metals ▶

DISPLAYED FORMULA

A displayed formula is a structural formula written out to show all the bonds in the structure. This is often called a *graphic formula*. The conventions used in writing displayed (graphic) formulae are shown in Fig. D.11.
◀ Formula ▶

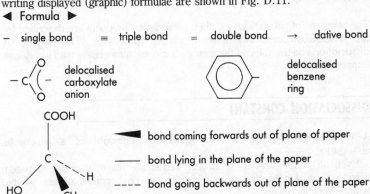

Fig. D.11 Conventions for displayed (graphic) formulae

DISPROPORTIONATION

A disproportionation reaction is a **redox reaction** in which the **oxidation number** of the same element in a reactant simultaneously decreases and increases. Disproportionation reactions occur with aqueous ions of **transition metals** in intermediate oxidation states and with the reaction of **halogens** with alkali; for example:

$$3MnO_4^{2-}(aq) + 4H^+(aq) \rightarrow \quad MnO_2(s) + 2MnO_4^-(aq) + 2H_2O(l)$$

$\uparrow$ $\qquad\qquad\qquad\qquad\quad \uparrow \qquad\qquad \uparrow$

$+6$ $\qquad\qquad\qquad\qquad$ down to $+4$ $\quad$ up to $+7$ $\quad$ [oxidation numbers]

$$Cl_2(aq) + \quad 2OH^-(aq) \rightarrow Cl^-(aq) + ClO^-(aq) + \quad H_2O(l)$$

$\uparrow$ $\qquad\qquad\qquad\qquad\qquad \uparrow \qquad\quad \uparrow$

0 $\qquad\qquad\qquad\qquad$ down to -1 $\quad$ up to $+1$ $\qquad$ [oxidation numbers]

◀ Redox reaction ▶

DISSOCIATION

Dissociation is the splitting of a molecule or ion into smaller molecules or ions. It is usually an **endothermic reversible** process, brought about by the action of heat upon substances or by the action of **polar solvents** upon weak electrolytes.

$$N_2O_4(g) \xrightarrow[]{heat} 2NO_2(g)$$

$$CaCO_3(s) \xrightarrow[]{heat} CaO(s) + CO_2(g)$$

$$CH_3CO_2H(aq) + H_2O(l) \rightleftharpoons CH_3CO_2^-(aq) + H_3O^+(aq)$$

The mixtures of substances and their dissociation products are described by the **dissociation constants** of the reversible processes. The **degree of dissociation** and the value of the dissociation constant increase with increasing temperature.

DISSOCIATION CONSTANT

A dissociation constant is the **equilibrium constant** of a reversible dissociation.

Even on gentle heating, dinitrogen tetraoxide dissociates into nitrogen dioxide to give a brown gaseous mixture which becomes darker as the temperature increases:

$$N_2O_4(g) \rightleftharpoons 2NO_2(g).$$

At about 80°C, the value of the dissociation constant, K_P, is about 4 atm:

$$\frac{P_{NO_2}^2}{P_{N_2O_4}} = 4 \text{ atm (at about 80°C)}$$

If α is the **degree of dissociation** of the 1 mol of N_2O_4 at 80°C, then the equilibrium mixture will contain $(1-\alpha)$ mol of N_2O_4 and 2α mol of NO_2; that is a total of $(1-\alpha) + 2\alpha = (1 + \alpha)$ mol of gas molecules. If the total equilibrium pressure is P atm, then the partial pressures of the N_2O_4 and NO_2 will be $P(1-\alpha)/(1+\alpha)$ atm and $P2\alpha/(1+\alpha)$ respectively. Substituting these values into the above equation and simplifying gives: $\alpha^2 = 1/(P+1)$. So, the degree of dissociation, α, is about 0.7 when $P = 1$ and 0.5 when $P = 3$. This decrease in the extent of dissociation of N_2O_4 with increasing total pressure is to be expected from the **law of chemical equilibrium** and **le Chatelier's principle**.

Calcium carbonate must be heated to nearly 1000°C before the dissociation constant reaches 4 atm:

$$CaCO_3(s) \rightleftharpoons CaO(s) + CO_2(g).$$

The value of the dissociation constant equals the pressure of carbon dioxide which is called the dissociation pressure of the calcium carbonate; that is:

$$P = 4 \text{ atm (at about 1000°C)}$$

The value of this pressure is unaffected by the amounts of solid calcium carbonate and calcium oxide present.

Weak acids such as ethanoic acid and **weak bases** such as ammonia only partially dissociate into ions in aqueous solution. This dissociation is also called the ionisation of electrolytes. The degree of dissociation, α, increases if the solution is diluted and is regarded as complete at infinite dilution. The acid and base dissociation constants (K_a and K_b) and the degree of dissociation are related in Ostwald's dilution law: *the degree of dissociation of a weak electrolyte is proportional to the square root of the dilution(V)*, or

$$\alpha = \sqrt{(K_aV)}$$

DISTILLATION

Distillation is a purification technique in which a volatile substance is separated from less volatile substances by evaporating it, condensing its vapour and collecting the distillate elsewhere. The method is used, for example, to produce distilled water free of dissolved salts and to recover ethoxyethane after use in **solvent extraction**.

◀ Fractional distillation, Steam distillation ▶

DISTRIBUTION COEFFICIENT

The distribution coefficient is the ratio of the **concentrations** of a solute dissolved in two immiscible solvents in contact with each other.

DISTRIBUTION LAW

A solute will dissolve in two immiscible solvents in contact with each other such that at equilibrium the ratio of the concentrations of the solute will be a constant at a constant temperature.

The law does not hold if the solute reacts with either solvent or if the solute is associated or dissociated in either solvent. The ratio of the concentrations is called the distribution or partition coefficient. Its value depends on the temperature and on the nature of the solute and immiscible solvents, but not on the volumes of the two solvents.

It follows from the distribution law that in solvent extraction it is more efficient to use several successive small separate portions of solvent than to use one large portion of solvent. For example, suppose 1,000 ml of an aqueous mixture contains 1.0 g of an important organic compound and the distribution coefficient for the compound between ethoxyethane and water is 10:1. Four successive treatments with separate 25 ml portions of ethoxyethane would extract 59 per cent of the organic compound, compared to 50 per cent which would be extracted by one 100 ml portion of ethoxyethane.

◀ Solvent extraction ▶

DOUBLE BOND

A double bond is a covalent bond formed between two atoms by the sharing of two pairs of electrons. It is shorter and stronger than a single bond, but longer and weaker than a triple bond between the same two atoms.

DYNAMIC EQUILIBRIUM

Dynamic equilibrium is a physical or chemical equilibrium in which opposing kinetic molecular processes are taking place at exactly the same balancing rates, to give a system with constant intensive properties.

◀ Equilibrium ▶

EDTA

EDTA stands for ethylenediaminetetraacetic acid. It is an important complexing agent, providing a quadruply charged **anion** capable of acting as a hexadentate chelate **ligand** (see Fig. L.4). The systematic name for EDTA is 1,2-bis[bis(carboxymethyl)amino]ethane. It is widely used as a chelating agent in titrimetric analysis of metal cations such as Ca^{2+} and Mg^{2+} in hard water samples. EDTA is added to hydrogen peroxide to stabilise it against decomposition because the EDTA forms very stable **complexes** with any trace of **transition metal ions** that would otherwise catalyse the decomposition of H_2O_2 into H_2O and O_2.

◄ Chelate complex ►

EFFUSION

Effusion is the escape of gas through a tiny hole or a porous material from a higher to a lower pressure region. By applying **Graham's law** of diffusion and measuring the time taken for a known volume of gas to effuse, the **relative molecular mass** of the gas can be determined.

◄ Graham's law ►

ELECTROCHEMICAL CELL

An electrochemical (or *galvanic*) cell is a means of converting into electrical energy some of the chemical energy liberated by a **redox reaction**. The cell is usually composed of two **half-cells**, each of which consists of a metal **electrode** dipping into an aqueous electrolyte (see Fig. E.1). A **salt bridge** makes an electrolytic connection between the two **aqueous** electrolytes. When the electrodes are wired together, a current flows as electrons move through the connecting wire and ions move in the electrolytes and salt bridge.

Reduction occurs at the electrode in the half-cell receiving electrons from the connecting wire. **Oxidation** occurs at the electrode in the other half-cell. Taken together, these two electrode reactions constitute the redox reaction of the electrochemical cell.

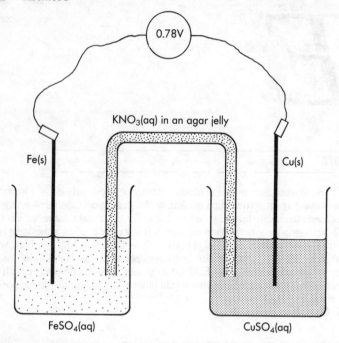

Fig. E.1

ELECTRODE

An electrode may be (a) a metal or graphite rod making electrical contact with the electrolyte in an **electrochemical cell** or in an electrolytic cell, (b) a metal plate and other solid chemicals forming a significant part of a battery or of an electrochemical **half-cell** or (c) an electrochemical half-cell.

Graphite electrodes are electrolytically inert. They do not take part in primary electrode reactions but they may be chemically attacked by, for example, oxygen released at the carbon anode surface by the primary electrode reaction during the commercial electrolytic extraction of aluminium. Platinum is usually inert but other metal anodes are much more likely to be attacked during electrolysis to form the cation. *Glass* electrodes are half-cells used for electrochemical analysis and for measurements of e.m.f. and pH. A standard **hydrogen electrode** is the half-cell used as the reference for standard **electrode potentials**.

ELECTRODE POTENTIAL

The electrode potential of a **half-cell** is the e.m.f. of a cell composed of that half-cell and a **hydrogen electrode**.

For example, the electrode potential of the $Cu^{2+}(aq) \mid Cu(s)$ electrode system is the e.m.f. of the **electrochemical cell** represented by the following **cell diagram**.

$$Pt[H_2(g)] \mid 2H^+(aq) :: Cu^{2+}(aq) \mid Cu(s)$$

The value of an electrode potential depends upon the temperature, pressure and **concentrations** of the electrolytes. The electrode potential is related to the standard electrode potential and the concentrations of the electrolytes by the **Nernst equation**.

▶ STANDARD ELECTRODE POTENTIAL

If the temperature is 298 K, the pressure is 1 atm and the electrolyte concentration is 1 mol dm^{-3} in both half-cells, then the copper is the positive electrode and the e.m.f. is 0.34 V, the standard electrode potential of the $Cu^{2+}(aq) \mid Cu(s)$ system.

The standard electrode potential (E^θ) is the e.m.f. of an electrochemical cell represented by a cell diagram in which a standard hydrogen electrode is shown as the left-hand half-cell and the conditions of temperature, pressure and electrolyte concentrations of the right-hand half-cell are standard. The sign given to the e.m.f. associated with the cell diagram of an electrochemical cell refers to the diagram's right-hand half-cell.

A table of standard electrode potentials constitutes an electrochemical series – see Fig. E.2. The values may be called standard reduction potentials when the half-cell electrode reactions are written as reductions – for instance: $Cu^{2+}(aq) + 2e^- \rightarrow Cu(s)$. The standard electrode potential for the hydrogen electrode is zero by definition because the e.m.f. of

$$Pt[H_2(g)] \mid 2H^+(aq) :: 2H^+(aq) \mid [H_2(g)]Pt$$

must be zero under standard conditions.

Combining standard electrode potentials

The value and sign of the standard e.m.f. of an electrochemical cell can be calculated from the standard electrode potentials of its two half-cells as follows:

For example, a $Cu^{2+}(aq) \mid Cu(s)$ half-cell ($E^\theta = +0.34$ V) andd a $Zn^{2+}(aq) \mid Zn(s)$ half-cell ($E^\theta = -0.76$ V) can be combined to form a Daniell cell:

$$Zn(s) \mid Zn^{2+}(aq) \; \vdots \; \vdots \; Cu^{2+}(aq) \mid Cu(s)$$

$$E^\theta_{cell} = (+0.34 \text{ V}) - (-0.76 \text{ V})$$

$$= +1.10 \text{ V}$$

The zinc electrode system is above the copper electrode system in the table of electrode potentials, so zinc will be the negative electrode and copper will be the positive. The negative (-0.76 V) zinc system may be thought to reppel **electrons** and the positive ($+0.34$ V) copper system to attract them. With one half-cell 'pushing' and the other 'pulling' electrons, the two half-cells are

ELECTRODE POTENTIAL

Right-hand half-cell	$E^{\ominus}/V$
$Li^+(aq) \mid Li(s)$	-3.03
$Rb^+(aq) \mid Rb(s)$	-2.93
$K^+(aq) \mid K(s)$	-2.92
$Ca^{2+}(aq) \mid Ca(s)$	-2.87
$Na^+(aq) \mid Na(s)$	-2.71
$Mg^{2+}(aq) \mid Mg(s)$	-2.37
$Al^{3+}(aq) \mid Al(s)$	-1.66
$Mn^{2+}(aq) \mid Mn(s)$	-1.19
$V^{2+}(aq) \mid V(s)$	-1.18
$[SO_4^{2-}(aq)+H_2O(l)], [SO_3^{2-}(aq)+2OH^-(aq)] \mid Pt$	-0.93
$Zn^{2+}(aq) \mid Zn(s)$	-0.76
$Cr^{3+}(aq) \mid Cr(s)$	-0.74
$Fe^{2+}(aq) \mid Fe(s)$	-0.44
$Cr^{3+}(aq), Cr^{2+}(aq) \mid Pt$	-0.41
$Cd^{2+}(aq) \mid Cd(s)$	-0.40
$V^{3+}(aq), V^{2+}(aq) \mid Pt$	-0.26
$Ni^{2+}(aq) \mid Ni(s)$	-0.25
$Co^{2+}(aq) \mid Co(s)$	-0.28
$Sn^{2+}(aq) \mid Sn(s)$	-0.14
$Pb^{2+}(aq) \mid Pb(s)$	-0.13
$2H^+(aq) \mid [H_2(g)]Pt$	± 0
$\frac{1}{2}S_4O_6^{2-}(aq), S_2O_3^{2-}(aq) \mid Pt$	$+0.09$
$[Sn^{4+}(aq)1.0M\ HCl], [Sn^{2+}(aq)(1.0M\ HCl)] \mid Pt$	$+0.15$
$Cu^{2+}(aq), Cu^+(aq) \mid Pt$	$+0.15$
$[4H^+(aq)+SO_4^{2-}(aq)], [H_2SO_3(aq)+H_2O(l)] \mid$	$+0.17$
$[PbO_2(s)+H_2O(l)], [PbO(s)+2OH^-(aq)] \mid Pt$	$+0.28$
$Cu^{2+}(aq) \mid Cu(s)$	$+0.34$
$[VO^{2+}(aq)+2H^+(aq)], [V^{3+}(aq)+H_2O(l)] \mid Pt$	$+0.34$
$Fe(CN)_6^{3-}(aq), Fe(CN)_6^{4-}(aq) \mid Pt$	$+0.36$
$[O_2(g)+2H_2O(l)], 4OH^-(aq) \mid Pt$	$+0.40$
$I_2(aq), 2I^-(aq) \mid Pt$	$+0.54$
$[MnO_4^{2-}(aq)+2H_2O(l)], [MnO_2(s)+4OH^-(aq)] \mid Pt$	$+0.59$
$[2H^+(aq)+O_2(g)], H_2O_2(aq) \mid Pt$	$+0.68$
$Fe^{3+}(aq), Fe^{2+}(aq) \mid Pt$	$+0.77$
$\frac{1}{2}Hg_2^{2+}(aq) \mid Hg(s)$	$+0.79$
$Ag^+(aq) \mid Ag(s)$	$+0.80$
$[2NO_3^-(aq)+4H^+(aq)], [N_2O_4(g)+2H_2O(l)] \mid Pt$	$+0.80$
$[ClO^-(aq)+H_2O(l)], [Cl^-(aq)+2OH^-(aq)] \mid Pt$	$+0.89$
$2Hg^{2+}(aq), Hg_2^{2+}(aq) \mid Pt$	$+0.92$
$[NO_3^-(aq)+3H^+(aq)], [HNO_2(aq)+H_2O(l)] \mid Pt$	$+0.94$
$[HNO_2(aq)+H^+(aq)], [NO(g)+H_2O(l)] \mid Pt$	$+0.99$
$[VO_2^+(aq)+2H^+(aq)], [VO^{2+}(aq)+H_2O(l)] \mid Pt$	$+1.00$
$Br_2(aq), 2Br^-(aq) \mid Pt$	$+1.09$
$[2IO_3^-(aq)+12H^+(aq)], [I_2(aq)+6H_2O(l)] \mid Pt$	$+1.19$
$[MnO_2(s)+4H^+(aq)], [Mn^{2+}(aq)+2H_2O(l)] \mid Pt$	$+1.23$
$[Cr_2O_7^{2-}(aq)+14H^+(aq)], [2Cr^{3+}(aq)+7H_2O(l)] \mid Pt$	$+1.33$
$Cl_2(g), 2Cl^-(aq) \mid Pt$	$+1.36$
$[PbO_2(s)+4H^+(aq)], [Pb^{2+}(aq)+4H_2O(l)] \mid Pt$	$+1.46$
$Mn^{3+}(aq), Mn^{2+}(aq) \mid Pt$	$+1.49$
$[MnO_4^-(aq)+8H^+(aq)], [Mn^{2+}(aq)+4H_2O(l)] \mid Pt$	$+1.51$
$[2HClO(aq)+2H^+(aq)], [Cl_2(g)+2H_2O(l)] \mid Pt$	$+1.63$
$[MnO_4^-(aq)+4H^+(aq)], [MnO_2(s)+2H_2O(l)] \mid Pt$	$+1.70$
$[H_2O_2(aq)+2H^+(aq)], 2H_2O(l) \mid Pt$	$+1.77$
$Co^{3+}(aq), Co^{2+}(aq) \mid Pt$	$+1.81$
$S_2O_8^{2-}(aq), 2SO_4^{2-}(aq) \mid Pt$	$+2.01$
$[O_3(g)+2H^+(aq)], [O_2(g)+H_2O(l)] \mid Pt$	$+2.08$
$F_2(g), 2F^-(aq) \mid Pt$	$+2.87$
$[F_2(g)+2H^+(aq)], 2HF(aq) \mid Pt$	$+3.06$

Fig. E.2 Standard electrode potentials

working together to move the electrons in the same direction, so the 0.76 and 0.34 add together to give 1.10 V.

If the standard electrode potentials of both half-cells have the same sign, they will be working in opposition to each other (both 'pushing' or both 'pulling' electrons), so the numerical E^θ values without their signs are subtracted to find the difference. For example, the standard electrode potential for $Ag^+(aq) \mid Ag(s)$ is $+0.80$ V so if this half-cell is combined with the copper half-cell, the E^θ_{cell} of the resulting electrochemical cell would be:

$$Cu(s) \mid Cu^{2+}(aq) \mid\mid Ag^+(aq) \mid Ag(s)$$

$$E^\theta_{cell} = (+0.80\,V) - (+0.34V)$$

$$E^\theta_{cell} = +0.46\ V$$

ELECTROLYSIS

Electrolysis is the decomposition of a substance by electricity, employing direct current electricity passed through an **aqueous** or molten **electrolyte** by means of metal or graphite **electrodes**. At the surface of the negative electrode (**cathode**), hydrogen and/or metals are formed by the discharge of the positive **ions** (cations) in the electrolyte – for example:

$$Cu^{2+}(aq) + 2e^- \rightarrow Cu(s)$$

At the surface of the positive electrode (**anode**), non-metals may be formed by the discharge of the negative ions (anions) in the electrolyte or cations may be formed by the anodic **oxidation** of the metal electrode – for example:

$$4OH^-(aq) \rightarrow 2H_2O(l) + O_2(g) + 4e^- \text{ or } Cu(s) \rightarrow Cu^{2+}(aq) + 2e^-$$

The current is due in the external circuit to the movement of **electrons** and in the electrolyte to the movement of the **anions** and **cations**.

Electrolysis is used on the industrial scale for the extraction of chlorine (from sodium chloride), aluminium (from bauxite), the alkali and alkaline earth metals (from their molten salts), for the refining of copper and for the electroplating of, for example, a metal alloy onto a printed circuit board.

ELECTROLYTE

An electrolyte is a substance which conducts an electric current when **aqueous** or molten and may undergo decomposition at the **electrodes**.

The flow of current in the electrolyte is the movement in opposite directions of the **anions** and **cations**. The most common electrolytes are molten salts and aqueous **acids**, **bases** and salts. Electrolytes are broadly classified as strong or weak according to their **degree of dissociation** when aqueous. Salts, **mineral acids** and alkali metal **hydroxides** are regarded as strong electrolytes which are completely ionised. Organic acids and bases are usually weak electrolytes which obey Ostwald's dilution law.

◀ Degree of dissociation ▶

ELECTROMOTIVE FORCE (e.m.f.)

The electromotive force of an **electrochemical cell** is the maximum potential difference (voltage) between the **electrodes**. The voltage of an electrochemical cell increases as the current drawn from the cell decreases. The maximum voltage (the e.m.f.) is reached when the current is zero. The e.m.f. is measured using a very high resistance voltmeter or a potentiometer circuit with a very sensitive galvanometer to detect minute currents (see Fig. E.3).

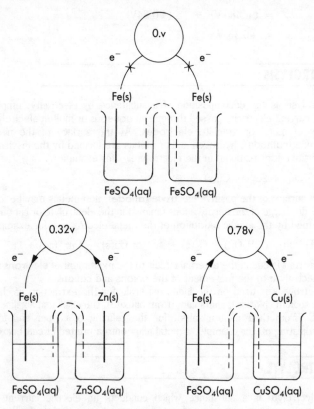

Fig. E.3 Direction of flow of electrons in electrochemical cells

The *standard e.m.f.* of an electrochemical cell is its maximum voltage at a temperature of 25°C, a pressure of 1 atm and with solutions of **concentration** equal to 1 mol dm^{-3}.

(For work beyond A-level, temperature, pressure and concentration are standardised at 298.15 K, 101.325 kPa and 1 mol kg^{-1} of solvent respectively.)

STANDARD e.m.f. AND GIBBS FREE ENERGY CHANGE

For the spontaneous **electrochemical cell** reaction, $\triangle G^\theta = -zFE^\theta_{cell}$ where z is the number of moles of **electrons** transferred per mole of **redox reaction** specified by the equation for the cell reaction and F is the **Faraday constant**.

In the **Daniell cell displacement reaction**, two electrons are transferred from a zinc **atom** to an **aqueous** copper(II) cation, so $z = 2$. The approximate value of F is 96,500 C mol^{-1}. Hence $\triangle G^\theta$ is $-2 \times 96,500 \times 1.10$J mol^{-1}, so:

$$Zn(s) + Cu^{2+}(aq) \rightarrow Zn^{2+}(aq) + Cu(s); \quad \triangle G^\theta = -212 \text{kJ mol}^{-1}.$$

This large negative value for $\triangle G^\theta$ means that the reaction is energetically very feasible. By contrast, the reverse reaction has an equally large positive value for $\triangle G^\theta$ and is not energetically feasible:

$$Cu(s) + Zn^{2+}(aq) \rightarrow Cu^{2+}(aq) + Zn(s); \quad \triangle G^\theta = +212 \text{kJ mol}^{-1}.$$

 ## STANDARD e.m.f. AND THE EQUILIBRIUM CONSTANT

E_{cell} $\triangle G^\theta$ and the equilibrium constant, K_c, for spontaneous cell reactions are interrelated:

$$\triangle G^\theta = -zFE^\theta_{cell} \text{ and } \triangle G^\theta = -RT \ln K_c \text{ so } E^\theta_{cell} = (RT/zF) \ln K_c$$

where F is the Faraday constant (96,500 coulombs per mole)
    R is the **gas constant** (8.31 joules per Kelvin per mole)
and T is the standard temperature (298 Kelvin)

Consequently, the standard e.m.f. of an electrochemical cell can be used to calculate a value for the equilibrium constant of the cell reaction. For the above displacement, $E^\theta_{cell} = +1.10$ V and $z = 2$. So,

$$1.10 = \frac{8.31 \times 298}{2 \times 96\,500} \times \ln K_c \text{ hence } \ln K_c = \frac{1.10 \times 2 \times 96\,500}{8.31 \times 298} = 85.7$$

This makes the value of K_c approximately 2×10^{37}. According to the **law of chemical equilibrium**,

$$K_c = \frac{[Zn^{2+}(aq)]}{[Cu^{2+}(aq)]} = 2 \times 10^{21}$$

This means the displacement goes to completion and there will be effectively no detectable amount of unreacted aqueous copper(II) cations in contact with metallic zinc.

ELECTRON

An electron (e$^-$) is a sub-atomic particle with a negative charge of approximately 1.602×10^{-19} C. The mass of an electron is about 1/1836 the mass of a **proton**. Electrons occupy the space around the **nucleus** of an atom and possess kinetic energy as a result of their movement and potential energy as a result of the attraction of the nucleus.

The patterns in the levels of energy of the electrons in the atoms may be related to the chemical and physical properties of the elements and their compounds. Electrons possess properties of both particles and waves. Free electrons emitted by radioactive decay of nuclei are known as **beta particles**.

ELECTRON AFFINITY

Electron affinity is the energy change when a **mole** of gaseous **atoms** or ions gain a mole of **electrons**. This energy change has been more precisely defined as the molar internal energy change, at 0 K, for processes such as:

$$X(g) + e^- \rightarrow X^-(g) \text{ and } X^-(g) + e^- \rightarrow X^{2-}(g).$$

More recently these processes have been called *electron gain* instead of electron affinity), and the energy change has been defined as the molar **enthalpy change**, at 298 K. For instance:

$$O(g) + e^- \rightarrow O^-(g); \qquad \triangle H_m^\theta \text{ (298 K)} = -141.4 \text{ kJ mol}^{-1}$$

$$O^-(g) + e^- \rightarrow O^{2-}(g); \qquad \triangle H_m^\theta \text{ (298 K)} = +790.8 \text{ kJ mol}^{-1}$$

The first electron gain enthalpies are usually negative values. The second electron gain enthalpies are positive because energy is required to overcome the repulsion of the electron by the negatively charged ion. Electron affinity must not be confused with **electronegativity**.

ELECTRONEGATIVITY

Electronegativity is a measure of how strongly an **atom** in a compound attracts **electrons** in a bond. The *Mulliken scale* of electronegativity is based on the molar first **ionisation** and the **electron affinity** of an atom. The *Pauling electronegativity index* (N_p) is based on calculations involving bond energies. Electronegativities increase from left to right across the periodic table and from the bottom to the top of the periodic table. So fluorine is the most electronegative of all the elements – see Fig. E.4.

H (2.1)		increasing electronegativity		
B (2.0)	C (2.5)	N (3.0)	O (3.5)	F (4.0)
	Si (1.8)	P (2.1)	S (2.5)	Cl (3.0)
		As (2.0)	Se (2.4)	Br (2.8)
			Te (2.1)	I (2.5)

Fig. E.4 Trends in the Pauling electronegativity index

Electronegativities can be used to predict the nature of the bonding between elements. The large difference between the electronegativity of a metal (low) and a non-metal (high) suggests **ionic bonding**. The small difference between the electronegativities of two non-metals suggests **covalent bonding**. The greater the difference in the electronegativities of two non-metal atoms, the more polar is the covalent bond between them.

◀ Intermediate bond, Polar molecule ▶

ELECTRONIC CONFIGURATION

The electronic configuration of an **atom** is a set of numbers, letters and superscripts representing the arrangement of the **electrons** into **orbitals** on the basis of their energies – for instance:

$$Pb\ 1s^2 2s^2 2p^6 3s^2 3p^6 3d^{10} 4s^2 4p^6 4d^{10} 4f^{14} 5s^2 5p^6 5d^{10} 6s^2 6p^2$$

The orbitals are grouped into shells (identified by the principal **quantum numbers** 1, 2, 3, etc.) and subshells (identified by the letters s, p, d and f). The superscripts show the number of electrons in each subshell. The s-subshell has one orbital, the p-subshell has three, the d-subshell has five and the f-subshell has seven orbitals. No more than two electrons may be in the same orbital so the maximum number of electrons in each of the first five levels of energy would be 2, 8, 18, 32 and 32 respectively. These numbers are related to the number of elements in each **period** across the **periodic table**: 2(H–He); 8(Li–Ne); 8(Na–Ar); 18(K–Kr); 18(Rb–Xe) and 32(Cs–Rn). The ground state electronic configuration of an element can be predicted from its **atomic number** using a plan of the periodic table and some simple rules; see Fig. E.5.

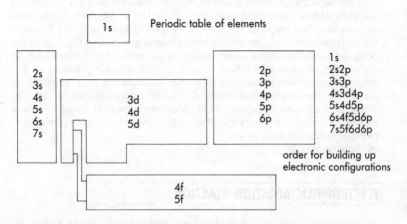

Fig. E.5 Order for building up electronic configurations

The electrons in the shell with the highest principal quantum number are referred to as the 'valence electrons in the outer shell'. These are the electrons with the lowest ionisation energies and the ones that are involved in bonding. The other electrons are the so-called 'inner shell electrons'. These are the electrons not usually involved in bonding. They partially shield the valence electrons (by repulsion between similar charges) from the attraction of the nucleus. Electrons in the same subshell do little to shield each other from the attraction of the nucleus.

Atoms of elements in the same group of the periodic table have the same number of valence electrons in the outer shell. This accounts for the similarity of properties of the elements in each group of the periodic table. The bonding in elements and their compounds can be related to the electronic configurations of the atoms. The stability of the atomic structure of the noble gases provides the basis of the octet rule.

ELECTRON PAIR REPULSION

Bonded and non-bonded pairs of electrons in the valence shell of an atom repel one another and these repulsions govern the shapes of molecules. For a given atom, the repulsion between two non-bonded pairs is greater than that between a bonded pair and a non-bonded pair; the repulsion is weakest between two bonded pairs.

◀ Valence shell electron pair repulsion theory (VSEPR) ▶

ELECTROPHILE

An electrophile is an electron-pair acceptor and a Lewis acid. NO_2^+, Br^+ and SO_3 are electrophiles which attack nucleophilic benzene and its derivatives in the electrophilic substitution reactions referred to as nitration, bromination and sulphonation respectively. In the reactions of aldehydes and ketones, the electronegative oxygen atom in the carbonyl group $(C=O)$ polarises the bond and makes the carbon atom an electrophile attacked by nucleophiles. In chloro- and bromoalkanes, the C–Cl and C–Br bonds are polarised by the electronegative halogen atoms so that the carbon atom is an electrophilic centre. In the alkaline hydrolysis of tertiary halogenalkanes by an S_N1 mechanism, the carbon–halogen bond undergoes heterolytic fission to form a carbo-cation in which the electrophilic centre is now a positively charged, electron-deficient carbon atom.

◀ Substitution ▶

ELECTROPHILIC ADDITION REACTION

An electrophilic addition is an addition reaction in which the first step is the attachment of an electrophile to a nucleophilic centre. Alkenes undergo electrophilic addition reactions because the high electron density of the $C=C$ bond makes the **molecules** nucleophilic. These additions follow the pattern of the following **reaction mechanism**:

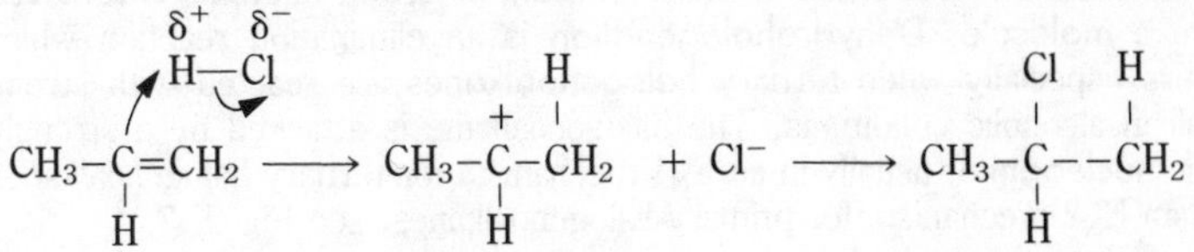

$$CH_3-\underset{H}{\overset{}{\underset{|}{C}}}=CH_2 \xrightarrow{\overset{\delta^+\ \delta^-}{H-Cl}} CH_3-\underset{H}{\overset{+}{\underset{|}{C}}}-CH_2 + Cl^- \longrightarrow CH_3-\underset{H}{\overset{Cl}{\underset{|}{C}}}--CH_2$$

propene   a secondary   2-chloropropane
                 carbocation

The addition occurs across the **double bond** and follows **Markovnikov's rule** to give 2-chloropropane and NOT 1-chloropropane as the product.

ELECTROPHILIC SUBSTITUTION

An electrophilic substitution (Fig. E.6) is a **substitution reaction** in which the attacking (and leaving) atom (or group) is an electron-pair acceptor or **electrophile**.

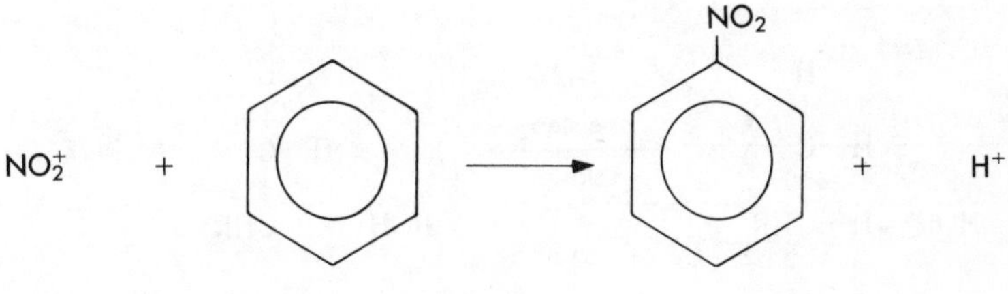

Fig. E.6

◀ Substitution ▶

ELEVATION OF THE BOILING POINT

The boiling point elevation is the raising of the temperature at which a liquid boils when a substance is dissolved in it. The rise ($\triangle t$) in the **boiling point** is a **colligative property** dependent upon the **concentration** (C_m) but not the nature of the solute particles according to the relationship $\triangle t = K_b C_m$. K_b (units K kg mol^{-1}) is the ebullioscopic constant for the liquid acting as a solvent. Measurement of the elevation of boiling point of a measured mass of solvent, of known K_b, by a measured mass of solute has been used to determine the **molar mass** of the solute.

◀ Colligative property ▶

ELIMINATION

An elimination is a reaction in which an **atom** or **group** of atoms is removed from a **molecule**. **Dehydrohalogenation** is an elimination reaction which occurs especially when tertiary **halogenoalkanes** are reacted with strong alkali in alcoholic conditions. The halogenoalkane is attacked by a strongly basic **nucleophile**, usually in an E_N1 mechanism for tertiary halogenoalkanes and an E_N2 mechanism for primary halogenoalkanes: see Fig. E.7.

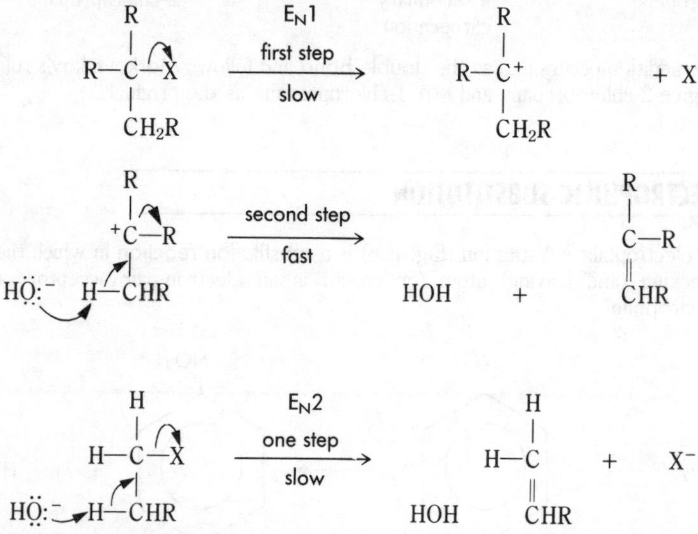

Fig. E.7 E_N1 and E_N2 reactions

A **condensation reaction** may be regarded as an **addition reaction** followed by an elimination reaction. **Dehydration** of an **alcohol** to an **alkene** may be regarded as an elimination in which the elements of water are removed from a molecule.

ELUENT

The eluent is the mobile phase carrying the mixture being analysed across or through the stationary phase in a chromatographic analysis. In the separation and identification of **amino acids** in a mixture by **paper chromatography**, the liquid eluent could be aqueous ethanoic acid in butan-1-ol. In the analysis of a mixture of alkanes by **gas chromatography** the eluent could be hydrogen or nitrogen gas. The eluent is sometimes called the *solvent*.

◀ Chromatography ▶

EMPIRICAL FORMULA

The empirical formula shows the composition of a substance as the simplest ratio of the amounts of the constituent elements. For example, the empirical formula of benzene is CH but the molecular formula is C_6H_6.
◀ Formula ▶

ENANTIOMER

An enantiomer is one of a pair of optical isomers whose molecular structure does not have a plane of symmetry.
◀ Optical isomerism ▶

ENDOTHERMIC REACTION

An endothermic reaction is a reaction in which heat energy is transferred from the surroundings to the system to raise the temperature of the products to the initial temperature of the reactants (Greek *endo* –inside; *thermos* – hot).
◀ Enthalpy change ▶

END-POINT

The end-point is the point in a titration revealed by a significant change in the property of the liquid system being monitored. This is usually shown by a sharp change in the colour of an indicator present in the liquid. For example, during the addition of an aqueous acid from a burette to a conical flask containing an aqueous base and bromothymol blue, the colour of the solution will change from blue to green, and finally to yellow. The appearance of the green colour marks the end-point of the titration and is taken to indicate the equivalence point in the acid–base reaction.

Indicators are the most common means of detecting end-points. However, other properties of the solution, apart from the colour caused by an added indicator, may be monitored to detect the end-point. The electrical conductance can be measured, or the solution can be incorporated into an electrochemical cell so that the e.m.f. of the cell can be measured. If the reaction is accompanied by a significant heat change, the temperature of the solution can be measured. In these cases it may be necessary to plot a graph of the physical property against cumulative volume of solution added from the burette in order to find the exact point at which the property changes – see Fig. E.8.
◀ Titration curves ▶

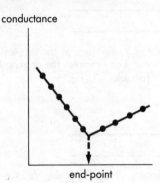

Fig. E.8 Finding the end-point in a titration

ENTHALPY CHANGE

An enthalpy change ($\triangle H$) is a heat change at constant pressure. In an **exothermic** process $\triangle H$ is negative – heat is 'taken away' by the surroundings to cool and restore the products to the initial temperature of the reactants. In an **endothermic** process $\triangle H$ is positive – heat is 'added' by the surroundings to warm and restore the products to the initial temperature of the reactants. (Greek *exo* – outside; *endo* – inside; *thermos* – hot). The majority of spontaneous chemical changes are exothermic.

 MOLAR ENTHALPY CHANGE

The heat change (δh) of a process is determined by measuring the temperature change (δT) it causes in a calorimeter and multiplying by the heat capacity (C) of the calorimeter and its contents: $\delta h = C \times \delta T$. This heat change which is for a measured but small amount of reaction is used to calculate the molar heat change. In some cases the term 'molar' refers to the 'amount of change' specified by the (chemical) equation representing the amounts of substances involved in the (chemical) change.

Enthalpy change ($\triangle H$) should not be confused with **internal energy change** (heat change at constant volume: $\triangle U$). $\triangle H = \triangle U + \triangle nRT$, where R is the gas constant, T is the absolute temperature and $\triangle n$ is the number of moles of gaseous products minus the number of moles of gaseous reactants. For example, the reaction of magnesium with hydrochloric acid releases more heat when it takes place at constant volume (in a sealed vessel) than when it takes place at constant pressure (in an open vessel). At constant pressure, the hydrogen uses up some of the heat of reaction in order to 'push back the atmosphere' as it escapes from the open vessel. A **flame calorimeter** measures the enthalpy of combustion ($\triangle H_C$) and a **bomb calorimeter** measures the internal energy of combustion ($\triangle U_C$). For A-level chemistry the emphasis is upon molar enthalpy changes.

Standard molar enthalpy changes

Since the value for the molar enthalpy change ($\triangle H$) of a process may vary with the temperature, pressure, concentration and physical state of the reactants and products, we usually standardise these conditions as follows.

$\triangle H^{\theta}_{298}$ represents a standard molar enthalpy change for a process in which all substances are in their most stable forms at a pressure of 1 atm (101 kPa) and a temperature of 298 K (25°C), and the **concentration** of any solution is 1 mol dm^{-3}.

Standard molar enthalpy change of combustion

$\triangle H^{\theta}_{c,298}$ represents the heat change at constant pressure when one **mole** of substance (at 298 K and 1 atm) is completely burnt in oxygen to form products (at 298 K and 1 atm). It is always negative.

For an organic compound such as $C_4H_9NH_2(l)$, complete combustion means the formation of $CO_2(g)$, $H_2O(l)$ and $N_2(g)$. Although combustion takes place above 298 K, the hot products are cooled down to 298 K after the combustion and all the heat 'taken away' by the surroundings constitutes the enthalpy of combustion. This includes the heat released when the water condenses to a liquid: $H_2O(g) \rightarrow H_2O(l)$; $\triangle H^{\theta} = -41$ kJ mol^{-1}.

Standard molar enthalpy change of atomisation

$\triangle H^{\theta}_{at,298}$ represents the heat change at constant pressure when one mole of gaseous **atoms** is formed from an element in its most stable form at 298 K and 1 atm. It is always positive.

Standard molar enthalpy change of formation

$\triangle H^{\theta}_{f,298}$ represents the heat change at constant pressure when one mole of a compound (at 298 K and 1 atm) is formed from its constituent elements in their most stable form at 298 K and 1 atm.

The standard enthalpies of formation of **oxides** are also the standard enthalpies of combustion of the elements. For example, the complete combustion of one mole of graphite is also the formation of one mole of carbon dioxide:

$$C(graphite) + O_2 \rightarrow CO_2(g); \quad \triangle H^{\theta}_{298} = -393.5 \text{ kJ mol}^{-1}$$

Standard molar enthalpy change of reaction

$\triangle H^{\theta}_{r,298}$ represents the standard enthalpy change for a reaction represented by a stoichiometric chemical equation (specifying the amounts of reactants and products involved in the reaction).

The standard molar enthalpy change of any reaction, real or imaginary, can be calculated from the standard molar enthalpies of formation of the reactants and products using **Hess's law**.

$$\triangle H^{\theta}_{r,298} = \Sigma \; \triangle H^{\theta}_{f,298}[\text{products}] - \Sigma \; \triangle H^{\theta}_{f,298}[\text{reactants}]$$

where Σ (Greek letter *sigma*) means the sum of all the molar enthalpies of formation, each multiplied by the appropriate stoichiometric coefficient.

ENTROPY

Entropy (S) is a measure of the random dispersal of energy of a system. The standard molar entropy (S^θ) of a substance at 298 K can be determined by measuring C_p (the molar heat capacity at constant pressure) at temperatures from 0 to 298 K, plotting a graph of a C_p/T (on the y-axis) against T (on the x-axis) and finding the area under the curve from 0 to 298 K. The units of entropy are J mol^{-1} K^{-1}.

When a substance turns from a solid into a liquid, its entropy increases because the particles can take up more energy as translational and rotational energy as well as vibrational energy. When the substance turns from a liquid into a gas, its entropy increases even more dramatically because the particles can take up so much more translational energy. The increase in the entropy of a substance on melting $(\triangle S_m)$ and on boiling $(\triangle S_b)$ may be calculated from the molar enthalpies of melting and boiling and the temperatures (in kelvin) at which these changes occur, thus:

$$\triangle S_m = \triangle H_m/T_m \text{ and } \triangle S_b = \triangle H_b/T_b$$

According to **Trouton's rule**, the standard molar entropy change of evaporation is constant for most liquids $(\triangle S_b \simeq 88 \text{ J K}^{-1} \text{ mol}^{-1})$.

The standard molar entropy change for a reaction may be calculated from the standard molar entropies of the reactants and products, thus:

$$\triangle S^\theta = \Sigma S^\theta(\text{products}) - \Sigma S^\theta(\text{reactants})$$

where Σ means the sum of the molar entropies, each multiplied by the stoichiometric coefficients in the chemical equation. For example,

$$\text{CoCl}_2.6\text{H}_2\text{O(s)} + 6\text{SOCl}_2\text{(l)} \rightarrow \text{CoCl}_2\text{(s)} + 12\text{HCl(g)} + 6\text{SO}_2\text{(g)}$$

| S^θ | 343 | 6×308 | 109 | 12×187 | 6×248 |
| | | | | | J mol^{-1} K^{-1} |

$$\triangle S^\theta \text{ is } 3841 - 2191 = +1650 \text{ J mol}^{-1} \text{ K}^{-1} (= 1.65 \text{ kJ mol}^{-1} \text{ K}^{-1})$$

Entropy increases when gases are formed and decreases when gases are removed.

According to the second law of thermodynamics, the total standard molar entropy change of a system and its surroundings always increases if a spontaneous chemical reaction occurs:

$$\triangle S^\theta_{\text{total}} = \triangle S^\theta_{\text{surroundings}} + \triangle S^\theta_{\text{system}}$$

and $\triangle G^\theta = \triangle H^\theta - T.\triangle S^\theta$

because $\triangle S^\theta_{\text{total}} = -\triangle G^\theta/T$ and $\triangle S^\theta_{\text{surroundings}} = -\triangle H^\theta/T$

If the $\triangle G^\theta$ is negative, then $\triangle S^\theta_{total}$ is positive and the reaction will be spontaneous (energetically feasible). $\triangle S^\theta_{system}$ must be large and positive for **endothermic reactions** to be energetically feasible. Endothermic reactions often give out a large volume of gas.

◀ Free energy change ▶

ENZYME

An enzyme is a **protein** formed in living cells that acts as a highly specific **catalyst** for one particular chemical reaction.

Enzymes consist of one of more **polypeptide** chains and have **molar masses** from about 10^4 to more than 10^6 g mol^{-1}. The specificity of enzyme catalysts depends upon the presence of **active sites** in the complex tertiary structure of these protein molecules. While the reaction is being catalysed, the **substrate** becomes attached to the enzyme at the active site to form a short-lived intermediate. Attachment can occur only if the substrate molecule has the correct structure to fit the enzyme shape at the active site. The tertiary structure of the enzyme and, therefore, its shape at the active site, is altered by changes in temperature and pH. Consequently, an enzyme has an optimum temperature and pH at which it functions most efficiently. Enzymes can be destroyed by high temperatures, which disrupt the tertiary structure and sometimes the secondary structure.

According to the type of reaction they catalyse, enzymes may be grouped into six broad classes: hydrolases, isomerases, ligases, lyases, oxidoreductases and transferases. The name of an enzyme is usually based on the name of the substrate and ends in -ase or -in. For example, saliva contains a group of enzymes called amylase which catalyse the **hydrolysis** of amylose (a constituent of the **polysaccharide** starch) into a mixture of glucose (a monosaccharide) and maltose (a disaccharide). Urease, an enzyme found in jack beans, catalyses the hydrolysis of **urea** but not closely related compounds such as thiourea $CO(NH_2)_2$:

$$CO(NH_2)_2(aq) + H_2O(l) \rightarrow NH_2CO_2OH(aq) + NH_3(aq)$$

◀ Proteins ▶

EQUILIBRIUM

Dynamic equilibrium (Latin *aequus* – equal; *libra* – balance) is the condition of constant intensive properties (such as **concentration**, pressure and temperature) achieved by a system of one or more substances when opposing kinetic molecular processes occur at exactly equal balancing rates.

Physical (or phase) equilibria are heterogeneous and involve changes of state of the same substances. They are important to an understanding of such techniques as **solvent extraction**, **steam distillation** and **fractional distillation**.

Chemical equilibria may be homogeneous (one phase) or heterogeneous (two or more phases) and involve reversible reactions. Important examples of reversible chemical reactions include the acid-catalysed hydrolysis of esters, the synthesis of hydrogen iodide and the industrial production of ammonia, nitric acid and sulphuric acid.

Application of the principles of equilibrium to aqueous acid-base and redox reactions is very important.

THE LAW OF CHEMICAL EQUILIBRIUM

If a reversible reaction represented by the chemical equation

$$aA + bB \rightleftharpoons cC + dD$$

is at equilibrium at a constant temperature, T, then

$$\frac{[C]^c \times [D]^d}{[A]^a \times [B]^b} \quad \text{has a constant value, } K_c,$$

and if the chemicals are all gaseous, then

$$\frac{P_C^c \times P_D^d}{P_A^a \times P_B^b} \quad \text{has a constant value, } K_p$$

where the letters A, B, C and D represent the chemical formulae of the reactants (A and B) and products (C and D) in the aqueous or liquid state and the letters a, b, c and d represent their stoichiometric coefficients in the balanced chemical equation. [A], [B], [C] and [D] represent the concentrations and P_A^a, P_B^b, P_C^c and P_D^d represent the partial pressures of the reactants and products.

EQUILIBRIUM CONSTANTS

The law of chemical equilibrium governs the chemical composition of a closed system of one or more reversible reactions at equilibrium at constant temperature. K_c and K_p are called the equilibrium constants because they represent the equilibrium constant values for the function of the concentrations or partial pressures. Note that a chemical equation is read from left (reactants) to right (products) and that in the expression for K_c (or K_p) the products are on the top (in the numerator) and the reactants are on the bottom (in the denominator). If the subscript (eq) is used to show that the concentrations or partial pressures have the equilibrium values, then at a constant temperature

$$K_c = \frac{[C]^c_{eq} \times [D]^d_{eq}}{[A]^a_{eq} \times [B]^b_{eq}}$$

$$K_p = \frac{P_{Ceq}^c \times P_{Deq}^d}{P_{Aeq}^a \times P_{Beq}^b} \qquad K_p = K_c \times (RT)^{(c+d-a-b)}$$

The value and the units of these equilibrium constants depend upon the balanced chemical equation for the reversible reaction, as can be seen from Fig. E.9. The value of K_c (or K_p) for a reaction is the reciprocal of the value of K_c (or K_p) for its reverse reaction, and the units alter accordingly. If the number of terms (raised to their appropriate powers) in the top (numerator) and bottom (denominator) of the equilibrium constant expression balance out, then $K_p = K_c$ and the value of the equilibrium constant is non-dimensional (has no units).

$T/°C$	Balanced chemical equation	K_p	Units	K_c	Units
80	$N_2O_4(g) \rightleftharpoons 2NO_2(g)$	4	atm	0.138	$mol\,dm^{-3}$
80	$\frac{1}{2}N_2O_4(g) \rightleftharpoons NO_2(g)$	2	$atm^{\frac{1}{2}}$	0.371	$mol^{\frac{1}{2}}\,dm^{-1\frac{1}{2}}$
80	$NO_2(g) \rightleftharpoons \frac{1}{2}N_2O_4(g)$	1/2	$atm^{-\frac{1}{2}}$	2.69	$mol^{-\frac{1}{2}}\,dm^{-1\frac{1}{2}}$
80	$2NO_2(g) \rightleftharpoons N_2O_4(g)$	1/4	atm^{-1}	7.25	$mol^{-1}\,dm^3$
830	$\frac{1}{2}H_2(g) + \frac{1}{2}I_2(g) \rightleftharpoons HI(g)$	5	–	5	–
830	$H_2(g) + I_2(g) \rightleftharpoons 2HI(g)$	25	–	25	–
830	$2HI(g) \rightleftharpoons H_2(g) + I_2(g)$	1/25	–	0.04	–

Fig. E.9 Chemical equations and equilibrium constants

▶ COMPOSITION

If the equilibrium concentration (or partial pressure) of one reactant is increased at constant temperature, the equilibrium concentrations of any other reactants decrease and those of the products increase.

This consequence of the law of chemical equilibrium has an important direct application: an inexpensive reactant can be used in excess of the amount required by the chemical equation to increase the percentage conversion of the other more expensive reactant(s) in the reversible reaction. For example, if one mole ethanoic acid is added to an equilibrium mixture (at 60°C) containing 0.33 mol ethanol and 0.67 mol ethyl ethanoate, there is a shift in the equilibrium composition. The amount of ethanol decreases to 0.15 mol and that of ethyl ethanoate increases to 0.85 mol. (In this example the amount of each substance is directly related to concentration because the volume term is the same for each substance and cancels.)

If the total number of moles of gaseous reactants is the same as that of the gaseous products, the equilibrium composition does not change with pressure. But if they are not the same, the equilibrium composition shifts in favour of the smaller number of moles with increasing pressure.

This important principle is illustrated by the fact that the **degree of dissociation** of hydrogen iodide ($2HI(g) = H_2(g) + I_2(g)$) does not change with pressure and that the **Haber process** for the production of ammonia ($N_2(g) + 3H_2(g) = 2NH_3(g)$) is carried out at 200 atm.

Le Chatelier's principle deals qualitatively with the way the composition of an equilibrium system may change with concentration and pressure.

▶ TEMPERATURE

When the temperature of any reversible reaction increases, the value of the equilibrium constant increases for the endothermic direction and decreases for the exothermic direction:

$$\ln(K_1/K_2) = (\triangle H^\theta/R)\{(1/T_2) - (1/T_1)\}$$

where $\triangle H^\theta$ is standard molar enthalpy change for the reaction, R is the **gas constant** and K_1 and K_2 are the equilibrium constant values at temperatures T_1 and T_2 kelvin respectively.

Since $\triangle H^\theta$ is positive for an **endothermic reaction**, if $\ln (K_1/K_2)$ is positive (because K_1 is greater than K_2), then T_1 must be greater than T_2 (to make $\{(1/T_2) - (1/T_1)\}$ positive). This fits the rule that K increases as the temperature of an endothermic reaction increases. An equilibrium mixture of nitrogen dioxide (brown) and dinitrogen tetraoxide (colourless) becomes darker brown in colour with increasing temperature because the following equilibrium 'shifts to the right' as the value of K_p increases:

$$N_2O_4(g) = 2NO_2(g); \quad \triangle H^\theta = +58 \text{ kJ mol}^{-1}$$

Le Chatelier's principle deals qualitatively with the way an equilibrium constant value may change with temperature.

▶ STANDARD e.m.f.

The standard e.m.f. (E^θ) of an **electrochemical cell** at the standard temperature (T) is related to the equilibrium constant (K) for the spontaneous cell reaction: $E^\theta = (RT/zF)\ln K$

▶ FREE ENERGY

The standard Gibbs free energy change ($\triangle G^\theta$) at the standard temperature (T) is related to the equilibrium constant (K) for the reaction:
$$\triangle G^\theta = -RT\ln K.$$

The reaction is energetically feasible if $\triangle G^\theta$ is negative. If $\triangle G^\theta$ is, say, less than -60 kJ mol^{-1}, the position of equilibrium is strongly in favour of products. If $\triangle G^\theta$ is, say, more than $+60$ kJ mol^{-1}, the position of equilibrium is strongly in favour of reactants.

EQUIVALENCE POINT

The equivalence point is the point in a titration at which the stoichiometric amounts of reactants have been added together. If 25.0 cm^3 of 0.100 mol dm^{-3} NaOH(aq) are titrated with 0.100 mol dm^{-3} HCl(aq) from a burette, the equivalence point will be reached when exactly 25.0 cm^3 of the **acid** have been added to the **alkali**. If a suitable indicator is used, the equivalence point will match the end-point of the **titration**.

◀ Titration curves ▶

ESTER

Name	Formula	Smell
ethyl methanoate	$HCO_2C_2H_5$	rum
3-methylbutyl ethanoate	$CH_3CO_2CH_2CH(CH_3)C_2H_5$	pear
methyl butanoate	$C_3H_7CO_2CH_3$	apple
ethyl butanoate	$C_3H_7CO_2C_2H_5$	pineapple
pentyl butanoate	$C_3H_7CO_2C_5H_{11}$	apricot
3-methylbutyl butanoate	$C_3H_7CO_2CH_2CH(CH_3)C_2H_5$	plum
octyl ethanoate	$CH_3CO_2C_8H_{17}$	orange
methyl 2-hydroxybenzoate	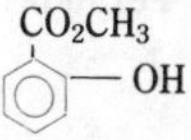 CO_2CH_3 ⬡—OH	wintergreen

Fig. E.10 Smells of some esters

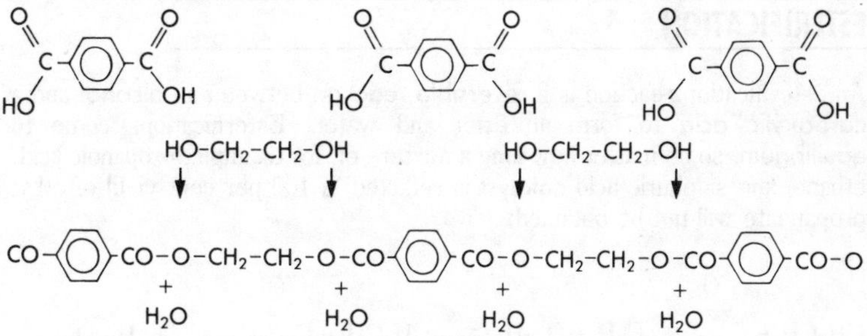

'Terylene' or 'Dacron' — linear polyester derived from ethane -1,2-diol

propane -1,2,3-triol
cross-linking three chains

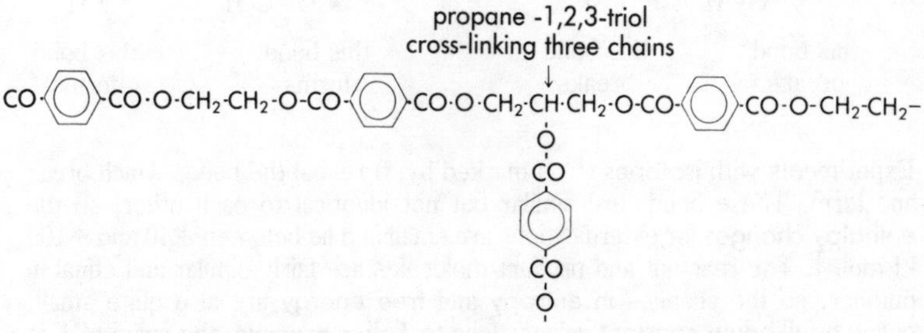

linear polyester chains (derived from ethane-1,2-diol) cross-linked into a three-dimensional structure by including propane-1,2,3-triol in the esterification process to produce an alkyd resin for use in paints

Fig. E.11 Polyesters of benzene-1,4-dicarboxylic acid

Esters are usually volatile liquids that are insoluble in water and identifiable by their strong fruity smells. Esters can be made by direct **esterification** of a **carboxylic acid** with an **alcohol** or by indirect esterification such as reacting an alcohol or a **phenol** with an **acyl chloride** and an **acid anhydride**.

Triglycerides (or triacylglycerols) are esters occurring naturally in plants and animals as unsaturated **oils** (such as olein) and saturated **fats** (such as stearin). They are esters of **long-chain aliphatic (fatty) acids** and propane-1,2,3-triol (glycerine or glycerol). Olein is an ester of octadec-9-enoic (oleic) acid and stearin is an ester of octadecanoic (stearic) acid. The **hydrogenation** of unsaturated oils to saturated fats using a nickel catalyst is the basis for making margarine – see Fig. H.7.

The most important reaction of esters is their **hydrolysis** to the carboxylic acid and alcohol (or phenol). **Refluxing** with **aqueous** sodium hydroxide hydrolyses an ester completely but produces the sodium salt of the carboxylic acid.

Esters are important industrial solvents. Many esters are synthesised industrially for use as food flavourings (see Fig. E.10). **Polyesters** are produced to make fibres for fabrics ('Terylene' or 'Dacron') and alkyd resins for paints (see Fig. E.11).

ESTERIFICATION

An esterification reaction is a **reversible reaction** between an **alcohol** and a **carboxylic acid** to form an **ester** and water. Esterifications come to **equilibrium**, so no matter how long a mixture of, for example, propanoic acid, ethanol and sulphuric acid **catalyst** is refluxed, a 100 per cent yield of ethyl propanoate will not be obtained:

$$CH_3CH_2-C\overset{O}{\underset{O-H}{\diagdown}} + C_2H_5-O-H \rightleftharpoons CH_3CH_2-C\overset{O}{\underset{O-C_2H_5}{\diagdown}} + H-O\diagdown H$$

| this bond breaks | this bond breaks | this bond forms | this bond forms |

Experiments with **isotopes** (^{18}O) marked by *) reveal the bonds which break and form. These bonds are similar but not identical to each other, so the **enthalpy changes** for esterifications are small and lie between $+40$ and $+100$ kJ mol^{-1}. The reactant and product molecules are fairly similar and equal in number, so the changes in **entropy** and **free energy** are also quite small, giving **equilibrium constant** values close to 1. For example, the value of K is about 4 for the esterification of ethanoic acid by ethanol.

Esterification reactions have high **activation energies** and are slow because strong bonds must break. However, the reaction can be catalysed by **mineral acids** – see Fig. E.12.

An acid catalyst could speed up the reaction by protonating the O-atom of the C=O in the carboxyl group:

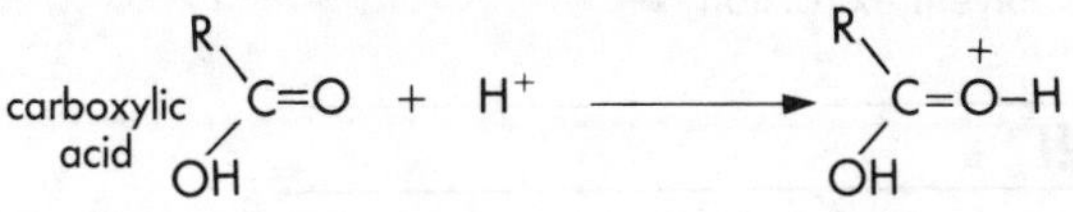

so that the nucleophilic O-atom of the OH in the alcohol can begin to form a bond before the C—O bond breaks:

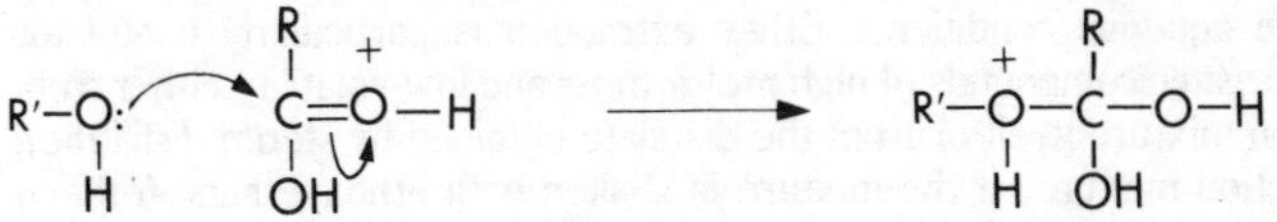

and then the OH can be eliminated as a water molecule and the H^+ catalyst regenerated:

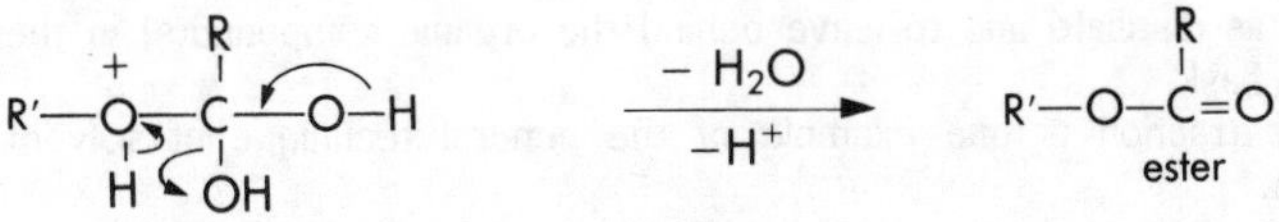

(Compare this mechanism with that for acyl chlorides on page 1)

Fig. E.12 Mechanism for acid-catalysed esterification

ETHANOYLATION

Ethanoylation is the introduction of a CH_3CO group into a molecule by reaction with ethanoyl chloride or ethanoic anydride.
◀ Acetylation ▶

ETHER

Ethers are **structural isomers** of **alcohols** and **phenols** in which the oxygen atom is connected to two carbon atoms – for example:

CH_3-O-CH_3	methoxymethane (dimethyl ether)
CH_3CH_2-OH	ethanol
$C_6H_5-O-CH_3$	methoxybenzene (anisole or methylphenyl ether)
$CH_3-C_6H_4-OH$	methylphenol(s)

Ethers are chemically rather unreactive but some, like ethoxyethane (diethyl ether or just 'ether'), have very low flash-points and ignite explosively. They can be prepared by **Williamson's synthesis**. They are used as industrial solvents. The low boiling point and immiscibility with water account for the use of ethoxyethane in the technique of solvent (ether) extraction.

◄ Ether extraction, Solvent extraction ►

ETHER EXTRACTION

The term ether extraction covers the use of ethoxyethane as an immiscible solvent to extract organic compounds from an aqueous mixture.

Preparation of organic compounds often involves reactions with inorganic reagents in aqueous conditions. Ether extraction is particularly useful for removing organic compounds of high molar mass and low volatility either from the reaction mixture itself or from the distillate obtained by steam distillation of the reaction mixture. If the mixture is shaken with ethoxyethane (ether), the organic compound(s) will dissolve in the ether layer, which is removed subsequently using a separating funnel. The inorganic reagents remain in the aqueous layer. The extraction may be repeated several times with small volumes of ether. The ether extracts are put together and distilled to recover the ether as distillate and to leave behind the organic compound(s) in the distillation flask.

Ether extraction is one example of the general technique of solvent extraction.

EUTETIC MIXTURE

A eutectic mixture is a solid mixture of constant composition that separates from a liquid mixture at a constant minimum freezing temperature below the freezing temperature of the separate pure components.

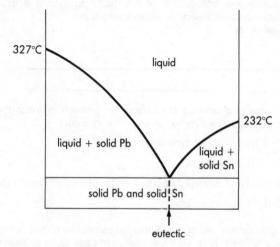

Fig. E.13 Phase diagram for mixture of lead and tin

One way in which eutectic mixtures can be obtained is from metals that form alloys. For example, tin (m.p. 232°C) and lead (m.p. 327°C) form a eutectic mixture with a **freezing point** of 183°C, the eutectic point, when the **mole fraction** of the tin is 0.75; see Fig. E.13. Sometimes the eutectic mixture is in the form of a solid solution in which there are no visible grains of the separate components.

EXOTHERMIC REACTION

An exothermic reaction is a reaction in which heat energy is transferred from the system to the surroundings to lower the temperature of the products to the initial temperature of the reactants (Greek *exo* – outside; *thermos* – hot).

◄ **Enthalpy change** ►

FACE-CENTRED CUBIC STRUCTURE

The face-centred cubic structure (Fig. F.1) is one of the two types of close-packed **crystal structure** adopted by metals. In both the face-centred cubic structure and the hexagonal structure the atoms have a **coordination number** of 12. The **close-packing** accounts for the hardness and high **density** of many metals compared to the soft, low-density **alkali metals** with their **body-centred structure**.

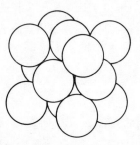

face-centred cubic

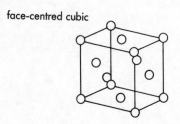

Fig. F.1

The ionic crystal structure of sodium chloride may be described as an interlocking **face-centred cubic structure**; see Fig. F.2. The sodium ions form a face-centred cube and so do the chloride ions. The **coordination number** of each ion is 6, so the ratio of the coordination numbers is 1:1 and the formula of the compound is Na^+Cl^-.

◀ Close-packing ▶

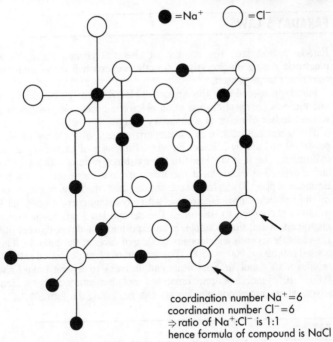

= Na$^+$ = Cl$^-$

coordination number Na$^+$=6
coordination number Cl$^-$=6
⇒ ratio of Na$^+$:Cl$^-$ is 1:1
hence formula of compound is NaCl

Fig. F.2 Sodium chloride face-centred cubic structure

FAJAN'S RULES

Covalent bonding is more probable and ionic bonding less probable if the ions possess multiple charges and if the atoms produce small **cations** *or large* **anions**.

Fajan's rules can be related to the **polarisation** of anions in terms of the **polarising power** of a cation and the **polarisability** of an anion. On the basis of these rules, lithium iodide, for example, would be expected to have more covalent character than caesium fluoride and less covalent character than aluminium iodide.

◀ Intermediate bond ▶

FARADAY CONSTANT

The Faraday constant, F, is the charge on one **mole** of **electrons** or one mole of **protons**. $F = eL$ where e (= 1.602×10^{-10} C) and L (= 6.022×10^{23} mol^{-1}) represent the elementary charge and the **Avogadro constant** respectively. The value of the Faraday constant is 9.648×10^4 C mol^{-1} but in A-level chemistry calculations is frequently approximated to 96,500 C mol^{-1}.

FARADAY'S LAWS

Faraday's first law: the amount of chemical change taking place at an electrode during electrolysis is directly proportional to the total amount of electrical charge passed.

Faraday's second law: the amount of electrical charge required to produce one mole of chemical change at an electrode during electrolysis is a simple whole number of moles of elementary charge.

The statements above are a modern version of the laws of electrolysis published in 1834 by Michael Faraday, Professor of Chemistry at the Royal Institution. The first law leads to the existence of ions with a constant mass and charge. For example, if the mass and charge on the metal cations in aqueous copper(II) sulphate were not constant, the mass of copper deposited on the cathode during electrolysis with a constant current need not increase in direct proportion to the time. The second law leads to the value of the charge on an ion. For example, in an experiment, a direct current of 0.500 A runs for 899 seconds and deposits 0.503 g of silver at the cathode. The charge passed is $0.500 \times 900 = 450$C. The amount of silver deposited is $0.503 \div 108 = 4.66 \times 10^{-3}$ mol Ag. So, the amount of charge to deposit 1 mol Ag would be 9.66×10^4 C mol^{-1} or approximately 1 mol elementary charges. Hence each silver cation carries one elementary charge, giving the formula Ag^+.

FAT

A fat is a naturally occurring ester (triglyceride) derived from a long-chain aliphatic (fatty) acid and propane-1,2,3-triol (glycerine).

Fats are solid at about 20° C and the hydrocarbon chain of the aliphatic acids is saturated – for instance:

$CH_3(CH_2)_{16}CO_2CH_2$
$CH_3(CH_2)_{16}CO_2CH$ stearin: an ester of octadecanoic (stearic) acid and
$CH_3(CH_2)_{16}CO_2CH_2$ propane–1,2,3-triol (glycerol)

Fats can be hydrolysed by boiling with concentrated aqueous alkali to produce glycerine and soaps.

◀ Oil, Triglyceride ▶

FATTY ACID

Fatty acids are straight-chain aliphatic carboxylic acids obtainable from animal fats (saturated) and vegetable oils (unsaturated). They usually contain an even number of carbon atoms and one carboxyl group on the end of the hydrocarbon chain. Octadecanoic (stearic) acid is a saturated fatty acid found in animal fat and octadec-9-enoic (oleic) acid is an unsaturated fatty acid found in olive oil. Both are long-chain acids of industrial importance in the production of soaps and dispersing agents.

f-BLOCK ELEMENTS

The f-block elements in the **periodic table** are the two horizontal rows of fourteen elements of atomic number 58 – 71 **(lanthanoids)** and 90 –103 **(actinoids)** following lanthanum and actinium respectively – see Fig. P.5.

FEHLING'S SOLUTION

Fehling's solution is a solution prepared by mixing equal volumes of **aqueous** copper(II) sulphate (Fehling's A) with alkaline aqueous sodium 2,3-dihydroxybutanedioate (tartrate) (Fehling's B).

This solution was first prepared by H.C. von Fehling, a German chemist, to test for **aldehydes** and reducing sugars. Although the blue solution is alkaline, copper(II) hydroxide does not precipitate because the Cu^{2+} ion is held in solution as a **complex** cuprate(II) **anion** with 2,3- dihydroxybutanedioate (tartrate) anions acting as bidentate ligands.

The sample under test is warmed with the solution and a positive result is indicated when the blue colour turns pale green (or even colourless) and an orange-red precipitate copper(I) oxide forms:

$$3CH_3-\underset{H}{\underset{|}{C}}=O + 2Cu^{2+}(aq) + 4OH^-(aq) \rightarrow$$

           blue solution
           copper(II)
           .complex

$$CH_3-\underset{OH}{\underset{|}{C}}=O + Cu_2O(s) + 2H_2O(l)$$

           orange-red precipitate
           copper(I) oxide

◀ Qualitative analysis ▶

FIRST-ORDER REACTION

A first-order reaction is a reaction with a differential **rate equation** of the form $-d[R]/dt = k[R]$, an integrated rate equation of the form $\ln R = -kt + \ln[R]_1$ and a constant **half-life** $t_{1/2} = k^{-1}\ln 2$, independent of **concentration**.

$[R]_1$ and $[R]$ represent the concentration of a reactant R initially (at $t = 0$) and at time t respectively. k is the **rate constant** $-d[R]/dt$, the rate of decrease of the concentration of a reactant with time, represents the **rate of reaction** which, in a first-order reaction, is directly proportional to the concentration raised to the first power – see Fig. F.3.

The decay of radioactive **isotopes** always follows first-order kinetics and the half-life is a characteristic property of a radioisotope. Decompositions of gases are often first-order reactions. For example, the reaction $N_2O_5(g) \rightarrow N_2O_4(g) + \frac{1}{2}O_2(g)$ has the following differential rate equation: $-d[N_2O_5(g)]/dt = k[N_2O_5(g)]$. Some reactions are called 'pseudo-first-order' reactions because a reactant which affects the rate is in such great excess that its concentration

$-d[\ ]/dt = k_1[\ ]^1$

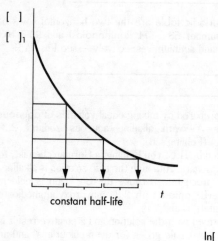

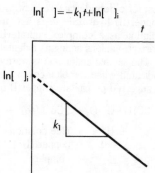

Fig. F.3 First-order reaction kinetics

remains effectively constant. For example, hydrolysis of sucrose to fructose and glucose is first-order with respect to sucrose and *appears* to be a first-order reaction overall. However, this is a pseudo-first-order reaction, because any change in the concentration of the water is negligible.

◄ Radioactivity ►

FLAME CALORIMETER

A flame calorimeter (Fig. F.4) measures heats of combustion at constant pressure. It is sometimes called a 'food' calorimeter because it may be used to measure the 'calorific value' of edible organic substances. The heat capacity of the calorimeter may be determined by finding the temperature rise produced either by the complete combustion of a measured mass of a substance whose molar enthalpy of combustion is known or by an electrical heater using a measured current at a measured voltage for a measured time. The flow of air and the height of the burner's wick should be adjusted to give a steady, smokeless flame and minimise incomplete combustion.

◄ Bomb calorimeter, Enthalpy change ►

FLAME TEST

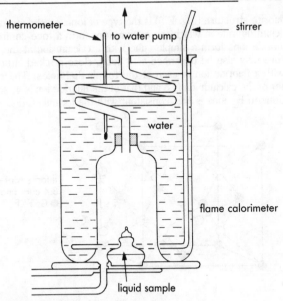

thermometer — to water pump — stirrer

water

flame calorimeter

liquid sample

Fig. F.4 Measuring heat of combustion at constant pressure

A flame test for the presence of certain metals is performed by moistening the sample with concentrated acid, placing it into a bunsen flame and observing the resulting colour of the flame. The hydrochloric acid produces **chlorides**, which are among the most volatile of the inorganic compounds. The test is used to detect the metals such as copper and lead and the **s-block metals** except beryllium and magnesium:

metal	flame colour
Li	scarlet
Na	yellow
K, Rb, Cs	lilac
Ca	brick red
Sr	scarlet
Ba	yellow-green
Cu	blue-green

Impurities of sodium compounds give a strong yellow flame colour which is persistent and often masks the colour due to the presence of elements such as potassium. Consequently, the colour of a flame should be observed through blue glass to filter out the yellow from the sodium. The lilac flame colour due to the presence of potassium compounds appears red when viewed through blue glass.

◀ Qualitative analysis ▶

FLUORITE STRUCTURE

The fluorite structure (Fig. F.5) is the type of **ionic crystal** structure adopted by calcium fluoride in which the calcium **ions** form a **face-centred cube** and the fluoride ions form a simple cube with a calcium ion at the centre. The structure may also be described as a **cubic close-packed** lattice of calcium ions with a fluoride ion in each of the tetrahedral holes. The **coordination number** of the calcium ion is 8 and that of the fluoride ion is 4, so the ratio of Ca^{2+} ions to F^- ions is 1:2, consistent with the formula CaF_2.

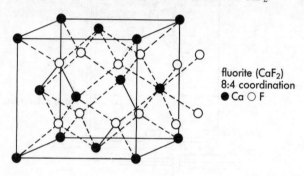

fluorite (CaF_2)
8:4 coordination
● Ca ○ F

Fig. F.5

◀ Calcite structure ▶

FORMULA

A formula is a collection of symbols showing the elements present in a substance – see Fig. F.6.

Empirical formula C_2H_4O	simplest possible formula
	Shows amounts of each element in a compound
	Deduced from composition by mass or from mass/charge ratio of parent molecule ion in mass spectrum
Molecular formula $C_4H_8O_2$	integral multiple of empirical formula
	Shows composition of molecule
	Deduced from empirical formula and relative molecular mass or from mass spectral analysis
Structural formula $(CH_3)_2CHCO_2H$	structure of a molecule
	Shows arrangement of atoms and groups in molecule
	Deduced from properties of a substance and from infra red absorption spectra and mass spectra

Displayed (graphic) formula	detailed structure of molecule
	Shows arrangement of all atoms and bonds in molecule as a projection of the structure onto a plane
	Deduced from structural formula by applying the rules of bonding to the atoms
Stereochemical formula	the spatial structure of a molecule
	Shows spatial arrangement of the bonds, atoms and groups in molecule as a 'perspective 3-D' formula
	Deduced from displayed formula by applying to the bonds, atoms and groups the principles of electron-pair repulsion and of electron delocalisation

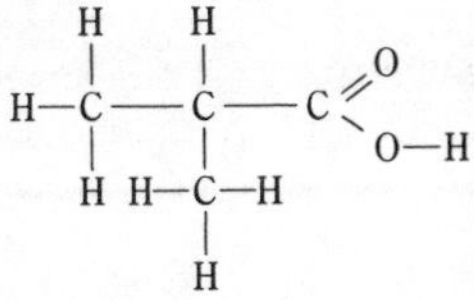

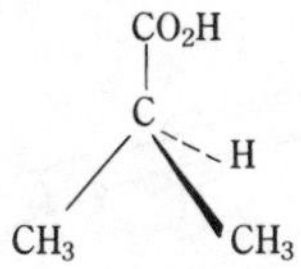

Fig. F.6 Types of formula for organic compounds

FRACTIONAL CRYSTALLISATION

Fractional crystallisation is a technique for separating a mixture of two or more soluble solids. The mixture is dissolved in the minimum quantity of a hot solvent and the solution allowed to cool slowly. The least soluble solute will crystallise first and may be removed from the solution containing the other more soluble solutes. Fractional crystallisation works well with solutes having significantly different solubilities, which vary considerably with temperature. It does *not* separate **eutectic** mixtures at their eutectic point.

FRACTIONAL DISTILLATION

Fractional distillation (Fig. F.7.) is a technique for separating a solution of two or more volatile miscible liquids. The technique works well with liquids which have significantly different boiling points and no tendency to form **azeotropes**. In the laboratory, a glass distillation apparatus is fitted with a fractionating column and the components of the solution are collected in succession, starting with the most volatile, as separate distillates in a receiver (see Fig. F.7). The technique is used in the **petrochemical** industry for separating crude oil into fractions, but these are continuously and simultaneously removed from the appropriate heights up the fractionating towers. The most volatile fraction is taken from the top of the tower and the progressively less volatile fractions are drawn from positions progressively lower down the tower.

◀ Distillation, Raoult's law ▶

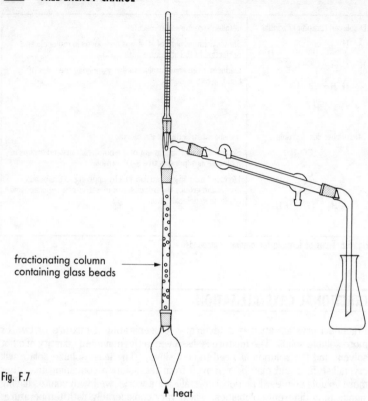

fractionating column
containing glass beads →

Fig. F.7

↑ heat

FREE ENERGY CHANGE

The standard (Gibbs) free energy change ($\triangle G^\theta$) for a process is the maximum energy, at 298 K and 1 atm pressure, available from that process to do useful work. $\triangle G^\theta$ is also a measure of the energetic feasibility of the process.

The standard free energy change of a **redox reaction** ($\triangle G^\theta$) can be calculated from the **standard e.m.f.** of an **electrochemical cell** having the required (redox) reaction as the cell reaction:

$$\triangle G^\theta = -zFE^\theta \text{ where } F \text{ is the Faraday constant,}$$

and from the **standard enthalpy** and **entropy changes** for the reaction:

$$\triangle G^\theta = \triangle H^\theta - T, \triangle S^\theta \text{ where } \triangle S^\theta \text{ is the standard molar entropy change.}$$

The standard free energy change of a reaction $\triangle G_r^\theta$ can be calculated as the difference between the sum of the standard free energies of formation of the products (each multiplied by the stoichiometric coefficients shown in the chemical equation) and that of the reactants

$$\triangle G_r^\theta = \Sigma \triangle G_f^\theta[\text{products}] - \Sigma \triangle G_f^\theta[\text{reactants}].$$

where Σ (Greek letter sigma) means the sum of all the molar free energies of formation, each multiplied by the appropriate stoichiometric coefficient.

$\triangle G_r^\theta$ must be negative if a reaction is to be spontaneous or energetically feasible. This rule comes from the second law of thermodynamics, which requires that the total entropy always increases, because $-\triangle G^\theta/T$ is $\triangle S_{total}^\theta$.

◄ Equilibrium constant, Standard electrode potential ►

FREE RADICAL

A free radical is a short-lived **atom** or **group** of atoms formed by the **homolytic fission** of a **covalent bond**. Free radicals are reactive because they contain an atom with an unpaired valence electron. In the **substitution reactions** of **alkanes** with **halogens** by a free **radical chain mechanism**, halogen radicals are formed in the initiation step; for example:

$$Cl_2 \xrightarrow{\text{light}} 2Cl\cdot$$

Free radicals are also involved in the **propagation** and **termination** steps of these reactions.

◄ Radical chain mechanism ►

FREQUENCY FACTOR

The frequency factor is the name of the constant, A, in the **Arrhenius equation** relating the **rate constant**, k, to the **activation energy**, E_a, of a reaction and the absolute temperature, T, at which the reaction takes place:

$$k = A. \, e^{-(E_a/RT)}$$

FREEZING POINT

The freezing point of a substance is the temperature at which the solid and liquid phases can coexist in **equilibrium** at a standard pressure of 1 atm. The freezing point is the same as the **melting point** of a substance but it is the term usually used when the substance is in the liquid phase and is being cooled down to form the solid phase.

Relative molecular masses of solutes have been determined by measurement of the **depression of the freezing point** of a liquid when a solute is added to it to form a solution. The *freezing point depression* is a **colligative property**.

FRIEDEL–CRAFT REACTION

A Friedel–Craft reaction is a type of **electrophilic substitution reaction** catalysed by aluminium chloride and used to introduce an alkyl or acyl group on to a benzene ring or into some other molecule. The reaction was discovered in 1877 by Charles Friedel, a French chemist, and James Craft, an American chemist, during research work together in France.

$$CH_3\overset{\displaystyle O}{\overset{\|}{C}}-Cl + AlCl_3 \rightarrow CH_3\overset{\displaystyle O}{\overset{\|}{C}}{}^+ + AlCl_4^-$$

$$C_6H_6 + CH_3\overset{+}{C}O \rightarrow C_6H_5\overset{\displaystyle O}{\overset{\|}{C}}-CH_3 + H^+$$

◀ Acylation, Alkylation ▶

FUNCTIONAL GROUP

A functional group is an element (such as Br) or combination of elements (such as OH) responsible for specific properties of an organic compound or class of compounds; see Fig. F.8.

Functional group		Prefix	Suffix	Class of compounds
>C:C<	$\diagdown C=C \diagup$		ene	alkenes
C_6H_5-		phenyl	benzene	arenes
—OH	—O—H	hydroxy	ol	alcohols and phenols
—CHO	$-C\diagup{}^{O}_{\diagdown H}$		al	aldehydes
>CO	$\diagdown C=O$	oxo	one	ketones and aldehydes
—CO$_2$H	$-C\diagup{}^{O}_{\diagdown O-H}$		oic acid	carboxylic acids
—CO$_2$—	$-C\diagup{}^{O}_{\diagdown O-}$		oate	esters and polyesters
—COCl	$-C\diagup{}^{O}_{\diagdown Cl}$		oyl chloride	acyl chlorides

Functional group		Prefix	Suffix	Class of compounds
$(-CO)_2O$ 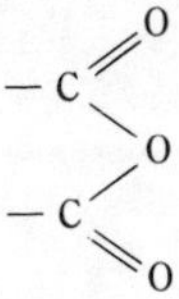	$\begin{array}{c}-C\diagup^{O}_{\diagdown O}\\\\-C\diagdown^{O}\end{array}$		anhydride	acid anhydrides
$-NH_2$ 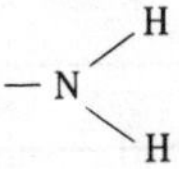	$-N\diagup^{H}_{\diagdown H}$	amino	amine	primary amines
$-CONH_2$ 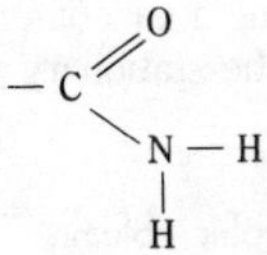	$-C\diagup^{O}_{\diagdown N-H}_{H}$		amide	amides
$-NO_2$ 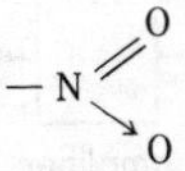	$-N\diagup^{O}_{\searrow O}$	nitro		nitro compounds
$-CN$	$-C\equiv N$	cyano	nitrile	nitriles
$-Hal$	$-Hal$	halogeno		halogeno compounds

Fig. F.8 Functional groups in organic compounds

GAS CHROMATOGRAPHY

Gas chromatography is a technique for separating a gaseous mixture using a gas as the mobile phase and a liquid or solid as the stationary phase (see Fig. G.1)

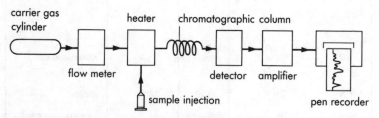

Fig. G.1 Arrangement for gas chromatography

In *gas-liquid chromatography* (GLC), the stationary phase is an involatile liquid film on a finely powdered solid support. In *gas–solid chromatography* (GSC), the stationary phase is just an inert porous solid. In either case, the stationary phase is usually packed into a long thin tube and the mixture of gases or vapours is carried through this column by hydrogen or nitrogen as the carrier gas. The components of the mixture separate because they have different coefficients of distribution between the mobile gas phase and the stationary liquid or solid phase. Those components which are strongly absorbed in the liquid or adsorbed on the solid pass through the column more slowly than those which are less strongly absorbed or adsorbed. Each component is detected as it leaves the column and identified by the time taken to pass through the column. The detection of a component leaving the column may be by a change either in the thermal conductivity of the issuing gas or in the ionisation of its flame. In sophisticated instruments, the components leaving the column may be analysed automatically by a **mass spectrometer**.
◀ Chromatography ▶

GAS CONSTANT

The gas constant, R, is the constant of proportionality in the equation of state for an ideal gas: $pV = nRT$.

R is a fundamental constant whose value to four significant figures in SI units is $8,314 \text{ J K}^{-1} \text{ mol}^{-1}$. There are at least another eighty-three different values of R, corresponding to combinations of the different units chosen for p, V, T and mass. $R = 0.08205 \text{ 1-atm K}^{-1} \text{ mol}^{-1}$ is a useful value to use in the **ideal gas equation** to calculate the molar mass of a gas from experimental measurements of its pressure, volume and temperature in the laboratory. In A-level examinations a value R will normally be provided as required.

R appears in three other important equations: the **Arrhenius equation**, relating the **rate constant** (k) to the temperature and **activation energy** (E_a) of a reaction; the **Nernst equation**, relating the potential of an electrode (E) to its **standard electrode potential** (E^{θ}) and the **concentration** of its ions; and the equation relating the **standard free energy change** $\triangle G^{\theta}$ of a reaction to its **equilibrium constant** (K).

◀ Ideal gas equation ▶

GAS LAWS

The gas laws are three laws (**Boyle's law, Charle's law** and the **pressure law**) connecting the pressure, volume and temperature of a fixed amount of (an ideal) gas. The **ideal gas equation** ($pV = nRT$) incorporates all three gas laws. Real gases approach ideal behaviour at high temperature and low pressures, and deviate most from ideal behaviour at temperature and pressures close to their point of liquefaction.

◀ Van der Waals' equation ▶

GAY-LUSSAC'S LAW

The volumes of gaseous reactants and products of a reaction, measured at the same temperature and pressure, will be in a simple ratio to each other.

Problem: 20.0 cm^3 of a gaseous hydrocarbon C_xH_y were exploded with 200.0 cm^3 of oxygen to produce 140 cm^3 of gaseous products which contracted to 40 cm^3 of oxygen after absorption with concentrated alkali. All volumes were measured at a temperature of 298K and a pressure of 1 atm. Find the value of x and y.

Solution:

$$C_xH_y(g) + (x + y/4)O_2(g) \rightarrow xCO_2(g) + (y/2)H_2O(l)$$

20 cm^3	$(200 - 40) \text{ cm}^3$	$(140 - 40) \text{ cm}^3$	$\simeq 0$ (liquid)
$\Rightarrow 20 \text{ cm}^3$	160 cm^3	100 cm^3	

applying Avogadro's principle to these volumes of gases gives the ratio of the amounts of gases (in moles):

$$1C_xH_y(g) + 8O_2(g) \rightarrow 5CO_2(g) + (y/2)H_2O(l)$$

$\Rightarrow$ x = 5 and $5O_2(g)$ is used to produce $5CO_2(g)$

leaving $3O_2$ (= 60) to produce $(y/2)H_2O(l)$, so $y/2 = 6$

$\Rightarrow$ y = 12 and the hydrocarbon is pentane, C_5H_{12}.

Fig. G.2 Combustion analysis of a hydrocarbon

The French chemist, Louis J. Gay-Lussac, put forward his law of combining volumes in 1809 to summarise the results of his experiments on reactions involving gases. The Italian chemist, Amadeo **Avogadro**, provided an explanation for this law with the **principle** he stated in 1811. Gay-Lussac's law and **Avagadro's principle** showed chemists how to determine formulae and equations by measuring the volumes of gaseous reactants and gaseous products – see Fig. G.2.

GEIGER–MARSDEN EXPERIMENTS

The Geiger–Marsden experiments measured the scattering of α-particles by metal foils and led to the nuclear model of the **atom**. At Manchester University in 1910, H. Geiger and E. Marsden found that a few **alpha particles** were scattered by a metal foil through angles greater than 90°. To account for this unexpected scattering, Professor Ernest Rutherford developed a mathematical expression which related the number of scattered α-particles to their velocity and angle of deflection and to the thickness of the metal foil and the charge on the supposed nucleus of its atoms.

Geiger and Marsden subsequently performed an elaborate series of α-particle scattering experiments which confirmed Rutherford's mathematical expression and his theory that an **atom** contains a minute but extremely dense, positively charged **nucleus**.

GEOMETRIC ISOMERISM

Geometric isomerism is a special form of **stereoisomerism** which occurs because two atoms joined by a **double bond** cannot rotate about the bond as an axis. The prefixes *cis-* (Latin – 'on this side') and *trans-* (Latin – 'on the other side') are used usually to name simply geometric isomers with the same **group** on each of two doubly bonded carbon atoms – for example:

cis-1,2-dichloroethene   *trans*-1,2-dichloroethene

GIANT STRUCTURES

A giant structure is a network of **atoms** or **ions** strongly held together in a three-dimensional lattice by **covalent, ionic** or **metallic bonding**. Substances with giant covalently bonded structures are usually poor electrical and thermal conductors with very high **melting points** and enthalpies of melting.

Diamond (an **allotrope** of carbon), silicon and silicon carbide (SiC) form similar giant molecular structures in which each atom is joined by covalent bonds to four other atoms located at the corners of a tetrahedron, giving an interconnected network of six-membered rings – see Fig. G.3. Graphite (another allotrope of carbon) forms a giant structure consisting of carbon atoms convalently bonded into large flat sheets of interconnected hexagons – see Fig. G.4. Within each sheet some delocalised **valence electrons** are mobile and give the structure its good electrical and thermal conductivity. Only weak non-directional **van der Waals forces** may attract the sheets to each other. The ability of the sheets to slide over each other gives the structure its soft, slippery lubricating property.

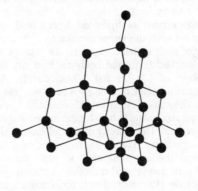

Fig. G.3 Structure of a diamond

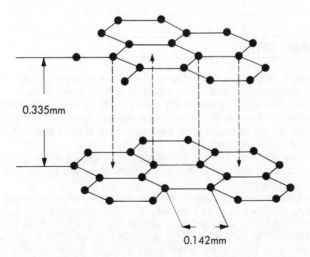

0.335mm

0.142mm

Fig. G.4 Structure of graphite

Substances with giant ionic structures are usually brittle, easily cleaved crystalline solids with high melting points and and molar enthalpies of melting. They are poor electrical and thermal conductors when solid but they conduct electrolytically when **aqueous** or molten. The structures of **caesium chloride**, sodium chloride and calcium fluoride (**fluorite**) are typical of ionic compounds. The exact type of ionic structure adopted by a compound is determined primarily by the relative sizes of the **cations** and **anions** together with the need for overall electrical neutrality and minimum repulsion between **ions** of like charge (usually the anions). Hydrated ionic crystals usually have lower melting points and molar enthalpies of melting than their anhydrous counterparts and are easier to cleave. The water molecules, held in the lattice by **non-dipole forces**, lower the electrostatic forces of attraction between the oppositely charged ions by keeping them further apart. In some cases the **crystal structure** may consist of **hydrated** layers held together by weak **hydrogen bonding** between the **water molecules** of hydration.

The three types of giant metallic structure are **hexagonal close-packed** (hcp), **cubic close-packed** (ccp) and **body-centred structures**; these have **coordination numbers** of 12, 12, and 8 respectively. All three may be described as a **lattice** of mutually repelling positive ions held together by their attraction for mobile delocalised electrons. This **metallic bonding** accounts for metals usually melting and boiling at high temperatures and consuming a great deal of heat in the process. The mobility of the delocalised electrons accounts for the lustre and conductivity. The strong non-directional attraction between the positive ions and the electrons also accounts for the hardness and toughness, coupled with ductility and malleability, shown by most metals and alloys. Most metals have close-packed structures and are usually harder and denser than the **alkali metals**, which have body-centred structures.

◀ Crystal structure ▶

GRAHAM'S LAW

At constant temperature and pressure, the rate of diffusion (or effusion) of a gas is *inversely* proportional to the square root of its **molar mass**.

Gaseous diffusion is the spread of a gas (or liquid) by the random movement of its **molecules** from a high to a low **concentration** region. Gaseous *effusion* is the escape of a gas through a small hole into a region of low concentration from a region of high concentration.

The time (t) taken for a fixed volume (V) of gas to effuse at constant temperature and pressure is *directly* proportional to the square root of its molar mass (M).

The molar masses $(M_1$ and $M_2)$ of the two gases may be compared by measuring the times $(t_1$ and $t_2)$ taken for the same volume of each gas under the same conditions of temperature and pressure to effuse from the same apparatus (see Fig. G.5). If you know the molar mass of one of the gases, you can calculate the molar mass of the other: $M_1/M_2 = (t_1/t_2)^2$.

◀ Kinetic theory ▶

graduated glass syringe

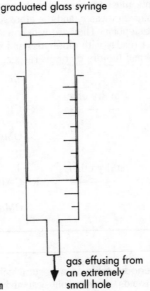

gas effusing from
an extremely
small hole

Fig. G.5 Graham's law of diffusion

GRAPHIC FORMULA

A graphic formula is another name for a displayed **structural formula** which
shows all the bonds in a **molecule** or **ion**.
◄ Formula ►

GRIGNARD COMPOUNDS

Grignard compounds are a class of **organo-metallic compounds** with the
general formula RMgX, where X represents Cl, Br or I and R represents an
alkyl or aryl group. These compounds were discovered by Victor Grignard, a
French chemist who won the Nobel prize in 1912. They are made by reacting
magnesium with organic halogen compounds in ethoxyethane under
completely anhydrous conditions. They are usually not separated but used in
the anhydrous ether as **Grignard reagents**.

GRIGNARD REAGENTS

A Grignard reagent is an organo-magnesium halide (Grignard compound) in
anhydrous ethoxyethane.

Grignard reagents must be handled with great care because they react highly exothermically with water, and the ethoxyethane is a highly flammable liquid with a low flash-point. They are widely used in organic **synthesis**. For example, they react readily with **aldehydes** and **ketones**, followed by water, to form **secondary** and **tertiary alcohols** respectively:

$$R'MgX + R\text{--}CHO \xrightarrow{\text{in dry ether}} \begin{array}{c} R' \\ | \\ R\text{--}CH \\ | \\ OMgX \end{array} \xrightarrow{\text{water}} \begin{array}{c} R' \\ | \\ R\text{--}CH \\ | \\ OH \end{array}$$

$$R'MgX + R_2CO \xrightarrow{\text{in dry ether}} \begin{array}{c} R' \\ | \\ R\text{--}C\text{--}R \\ | \\ OMgX \end{array} \xrightarrow{\text{water}} \begin{array}{c} R' \\ | \\ R\text{--}C\text{--}R \\ | \\ OH \end{array}$$

GROUP

A group in the **periodic table** is a vertical column of elements showing simimlarities and trends in their physical and chemical properties. The similarity of properties is explained by the similarity of the **electronic configurations**. For example, in Group I, the outer shell of the **alkali metal** atoms consists on one s-orbital electron. The trend in properties can be related to the increase in **atomic radius** with increasing **atomic number** and the shielding of the outer (valence) electrons by the complete inner electron shells.

A group is a combination of two or more covalently bonded elements responsible for specific properties of a compound or class of compounds.
◀ Functional group ▶

HABER PROCESS

The Haber process is the industrial manufacture of ammonia from nitrogen and hydrogen using a promoted iron **catalyst**. The mixture of nitrogen and hydrogen (synthesis gas) in the ratio 3:1 $H_2:N_2$ by mol is made chiefly from air (4:1 $N_2:O_2$ by mol), natural gas (CH_4) and some naphtha, a petroleum fraction. The **synthesis** of the ammonia depends upon the following **reversible reaction**:

$$N_2(g) + 3H_2(g) \quad \underset{\substack{\text{catalyst of iron} \\ \text{promoted by KOH}}}{\overset{\text{450°C 200 atm}}{\rightleftharpoons}} \quad 2NH_3(g); \quad \triangle H^\theta = -92 \text{ kJ mol}^{-1}$$

The conditions for achieving the optimum yield are dictated by the **law of chemical equilibrium** and the principles of reaction kinetics. When the gases are passed over the **heterogeneous catalyst** kept hot by the heat of the reaction, a conversion of about 15 per cent is achieved. The ammonia is liquefied by cooling and the 85 per cent unreacted synthesis gas is recycled.

The Haber process is also a source of sulphur and argon. Sulphur compounds are removed from the naphtha to avoid poisoning the catalysts and used as one source of sulphur for the manufacture of sulphuric acid. Atmospheric argon accumulates in the nitrogen–hydrogen mixture and has to be regularly removed to avoid the adverse effect of its increasing *partial pressure* upon the yield of ammonia.

◄ Contact process ►

HALF-CELL

A half-cell consists of a metallic **electrode** dipping into an **aqueous electrolyte**. The simplest half-cell is a metal dipping into its aqueous cations: for instance, copper dipping into aqueous copper(II) sulphate. It can be represented by the following half-cell diagram: ¦ Cu^{2+} aq | Cu(s). In these half-cells the electrode is not inert but participates in the half-cell reaction:

$$Cu^{2+}(aq) + 2e- \rightleftharpoons Cu(s).$$

In two other types of half-cell, the electrode is an inert metal, such as platinum. In one type, a shiny platinum electrode dips into an aqueous solution containing all the reactants and products of the half-cell reaction: ¦ Fe^{3+}(aq), Fe^{2+}(aq) | Pt. In the other type, a platinum electrode coated in platinum black dips into an aqueous electrolyte and also makes contact with a gas bubbling through the solution: ¦ $2H^+$(aq) | H_2(g) | Pt. Half-cells, often just called electrodes, are combined in pairs to give complete **electrochemical cells**.

◀ Electrochemical cell, Electrode, Hydrogen electrode, Standard electrode ▶

HALF-EQUATION

A half-equation is an equation for a half-reaction. The equation for an electron-transfer or a **proton-transfer** reaction may be regarded as the sum of two half-equations. For example, the equation for the **displacement** of copper by zinc may be written as a half-equation for the **reduction** of copper(II) **cations** and a half-equation for the **oxidation** of zinc **atoms**:

$$Zn(s) + Cu^{2+}(aq) \rightarrow Zn^{2+}(aq) + Cu(s)$$
$$Cu^{2+}(aq) + 2e^- \rightarrow Cu(s)$$
$$Zn(s) \rightarrow Zn^{2+}(aq) + 2e^-$$

Similarly, the equation for the reaction of hydrogen chloride with ammonia to form ammonium chloride may be written as a half-equation for the loss of a **proton** by the **acid** and a half-equation for the gain of a proton by the **base**:

$$HCl(g) + NH_3(g) \rightarrow NH_4Cl(s)$$
$$HCl(g) \rightarrow H^+ + Cl^-$$
$$NH_3(g) + H^+ \rightarrow NH_4^+$$

◀ Electrochemical cell ▶

HALF-LIFE

The half-life period of a process is the time taken for the **concentration** of a specified reactant to fall to half of its initial value. For a specific **first-order reaction** at constant temperature, this half-life period has a constant value which is independent of the concentration of the reactant.

The time taken for the **radioactivity** of a radio-isotope to fall to half of its initial value is an important characteristic of the **isotope**, being independent of the mass and unaffected by catalysts or by changes in temperature.

◀ Rate of reaction ▶

HALOGEN

The halogens are the elements (fluorine, chlorine, bromine, iodine and astatine) which make up Group VII in the **p-block** of the **periodic table**.

Electronic configuration	m.pt. /°C	b.pt. /°C	N_p	$E^\ominus/V$	Hydrides	Oxides
F [He]$2s^2 2p^5$	−220	−188	4.0	+2.87	H_2F_2 weak acid	OF_2
Cl [Ne]$3s^2 3p^5$	−101	−35	3.0	+1.36	HCl strong acid	Cl_2O; ClO_2; Cl_2O_7
Br [Ar]$3d^{10} 4s^2 4p^5$	−7	59	2.8	+1.09	HBr strong acid	Br_2O; BrO_2
I [Kr]$4d^{10} 5s^2 5p^5$	114	184	2.5	+0.54	HI strong acid	IO_2; I_2O_5

N_p = Pauling's electronegativity index

Fig. H.1 Properties of the halogens

 PHYSICAL PROPERTIES

The similarities between the elements are greater in this group than in any other p-block group. The halogens are typical non-metals. They exist as **diatomic covalent molecules**. Their **melting points, boiling points** and enthalpies of melting and boiling increase with **atomic number** down the group as the intermolecular **van der Waals forces** increase with the number of **electrons** in the **atoms** (see Fig. H.1). Chlorine, bromine and iodine are slightly soluble in water.

 CHEMICAL PROPERTIES

The outer p-subshell of the halogen atoms is just one electron short of the configuration of the neighbouring noble gas atoms. By gaining electrons from metals or by sharing electrons with less electronegative non- metals, halogens act as **oxidising agents** and complete their outer p-subshell. The **electronegativity** and the reactivity of the halogens decrease with atomic number down the group from fluorine to iodine. Fluorine is the most electronegative of all the elements, so its oxidation number is −1 in all its compounds. The other three halogens form compounds with **oxidation numbers** ranging from −1 to +7.

Halogens combine with most metals to form halides which are often largely ionic in character. The reactivity of the halogens decreases down the group with increasing atomic number so iodine oxidises iron to iron(II) iodide (FeI_2) but chlorine can oxidise iron to iron(III) chloride ($FeCl_3$) and can displace bromine and iodine from their **anions**. The halide anions have the same **electronic configurations** as the corresponding **noble gas** atoms; for example:

$$Cl(2,8,7) + e^- \rightarrow Cl^-(2,8,8) \quad [Ar\ 2,8,8]$$

However, iodides are more covalent in character than fluorides, owing to the increasing **polarisability** of the halide anion with increasing anionic radius. The alkali metal halides are ionic and soluble in water. The silver halides become increasingly covalent in character and less soluble in water down the group from AgF to AgI. Silver fluoride forms **hydrates**! Precipitation of AgCl, AgBr and AgI using aqueous silver nitrate, together with nitric acid and ammonia, is the basis of a practical test for halides. (Fig. H.2).

	$Cl^-(aq)$	$Br^-(aq)$	$I^-(aq)$
add $HNO_3(aq)$ and $AgNO_3(aq)$	white ppt. AgCl darkens in sunlight	off-white ppt. AgBr darkens in sunlight	yellow ppt. AgI stays yellow
add $NH_3(aq)$ to ppt.	ppt. dissolves completely	ppt. partially dissolves	ppt. unaffected
add $Cl_2(aq)$ and $CCl_3CH_3(l)$	aq. soln. colourless org. liq. colourless	aq. soln. brown org. liq. red-brown	aq. soln. brown org. liq. purple
add $Br_2(aq)$ and $CCl_3CH_3(l)$	aq. soln. brown org. liq. red-brown	aq. soln. brown org. liq. red-brown	aq. soln. brown org. liq. purple

$HNO_3(aq)$ allows only silver halides to precipitate

$NH_3(aq)$ forms $[Ag(NH_3)_2]^+(aq)$ and lowers the $Ag^+(aq)$ concentration enough to prevent the precipitation of any AgCl and to allow the precipitation of some of the AgBr and almost all of the AgI.

Fig. H.2 Testing for halides

Halogens form covalent compounds with non-metals including hydrogen, oxygen and even themselves: for instance, iodine monochloride (ICl) and iodine trichloride (ICl_3). The hydrogen halides are covalently bonded diatomic molecules with the general formula HX. They can be prepared in the laboratory by warming anhydrous phosphoric acid with potassium halide:

$$H_3PO_4(l) + NaX(s) \xrightarrow{\text{heat}} NaH_2PO_4(s) + HX(g)$$

The HX **bond length** increases, the **bond energy** decreases and the stability of the hydrides decreases with increasing atomic number of the halogen down the group. Hydrogen halides give misty fumes in damp air and dissolve rapidly in water, giving acids that can react with **alkalis**, basic **oxides**, **carbonates** and hydrogencarbonates to produce the halide salts (the word halogen means 'salt-former'). Intermolecular hydrogen bonding caused by the high electronegativity of fluorine gives hydrogen fluoride an exceptionally high boiling point and makes hydrofluoric acid an atypically **weak acid**. Hydrogen halides (HX) and ammonia ($:NH_3$) react to form a white smoke of ammonium halides (NH_4X):

$$X-H(g) + :NH_3(g) \xrightarrow[\text{needed as catalyst}]{\text{trace of moisture}} X^-[H:NH_3]^+(s)$$

Halogen oxides are covalent, acidic and, with the exception of iodine(V) oxide, mostly unstable and often dangerously explosive. Fluorine does not form **oxoacids**. Apart from iodic(V) acid, the oxoacids of the other halogens are mostly unstable and found only in aqueous solution. Oxoacids and oxosalts are usually formed by the **disproportionation** of a halogen in water and alkali respectively:

$$Cl_2(g) + 3H_2O(l) \rightarrow 2H_3O^+(aq) + Cl^-(aq) + ClO^-(aq)$$

Oxoacids and oxosalts may disproportionate if the **oxidation number** of the halogen is positive but below the maximum possible value. Chlorate(V) forms in hot alkali because the chlorate(I) disproportionates:

$$3ClO^-(aq) \rightarrow 2Cl^-(aq) + ClO_3^-(aq)$$

When gently heated just above its melting point, potassium chlorate(V) ($KClO_3$) disproportionates into potassium chloride (KCl) and potassium chlorate(VII) ($KClO_4$) which decomposes on stronger heating into potassium chloride and oxygen. Oxoacids and oxosalts of halogens are generally good **oxidising agents**.

The halogens are usually obtained by oxidation of the halide. The technically difficult industrial production of fluorine involves anodic oxidation of the F^- ion by the electrolysis of an anhydrous liquid mixture of KF and HF at 100°C: $2F^- \rightarrow F_2 + 2e^-$. The commercial production of chlorine is by **electrolysis** of brine, but in the laboratory chemical oxidising agents such as manganese(IV) oxide or acidified manganate(VII) ions are used to prepare chlorine gas from concentrated hydrochloric acid.

$$MnO_4^-(aq) + 8H^+(aq) + 5Cl^-(aq) \rightarrow Mn^{2+}(aq) + 4H_2O(l) + 2\tfrac{1}{2}Cl_2(g)$$

Chlorine is the oxidising agent used to displace bromine from aqueous bromide ions in sea water:

$$Cl_2(g) + 2Br^-(aq) \rightarrow 2Cl^-(aq) + Br_2(aq).$$

The commercial extraction of iodine involves partial **reduction** of iodate(V) to iodide followed by a reaction which is the reverse of disproportionation:

$$5I^-(aq) + IO_3^-(aq) + 6H^+(aq) \rightarrow 3I_2 + 3H_2O(l).$$

HALOGENATION

Halogenation is the introduction of one or more **halogen atoms** into the **molecule** of a compound. Fluorination, chlorination and bromination of **saturated hydrocarbons** may be carried out by direct reaction with the **halogen** as, for example, in the **free-radical substitution reaction** between clorine and methane:

$$CH_4 + Cl_2(g) \rightarrow CH_3Cl(g) + HCl(g)$$

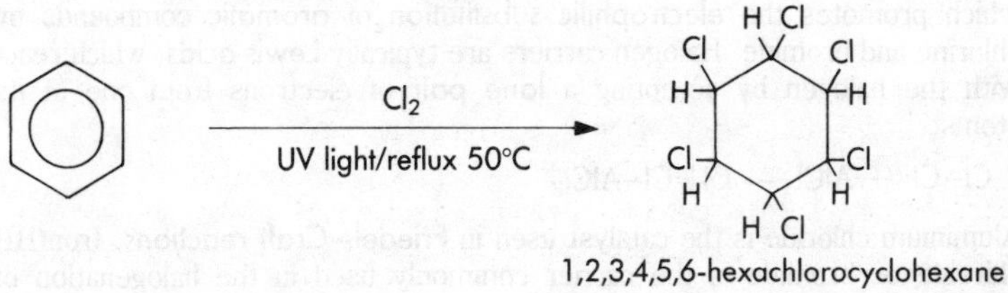

1,2,3,4,5,6-hexachlorocyclohexane

Fig. H.3 Chlorination of benzene

Benzene undergoes radical chain addition reactions with chlorine in ultraviolet light (or bright sunshine) ultimately to form a mixture of the nine isomers of benzene hexachloride with different arrangements of the H and Cl atoms above and below the ring. One of these isomers, called gamma-BHC, is an active insecticide sold commercially under the trade name 'Gammexane' (see Fig. H.3). Under similar conditions, methylbenzene undergoes radical chain substitution reactions in which the substituent methyl group is chlorinated with the formation of (chloromethyl)benzene, (dichloromethyl)benzene and (trichloromethyl)benzene:

$$C_6H_5-CH_3 \rightarrow C_6H_5-CH_2Cl \rightarrow C_6H_5-CHCl_2 \rightarrow C_6H_5-CCl_3$$
$$+ HCl \qquad + HCl \qquad + HCl$$

When an aluminium or iron(III) halide is used as a **halogen carrier**, electrophilic halogenation of the benzene ring occurs:

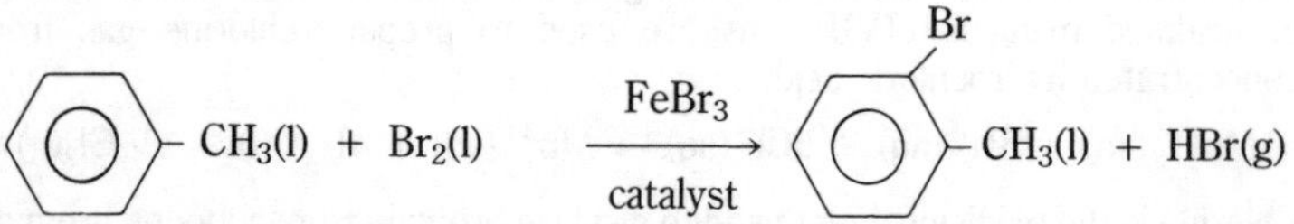

Halogenating agents such as the phosphorus halides or sulphur chloride oxide (thionyl chloride) can be used to attack OH groups in molecules, for instance:

$$CH_3OH(l) + SOCl_2(l) \rightarrow CH_3Cl(l) + SO_2(g) + HCl(g)$$

Iodine is too unreactive for direct iodination but alcohols can be reacted with a mixture of red phosphorus and iodine to produce iodoalkanes; for instance:

$$6C_2H_5OH(l) + 2P(s) + 3I_2(S) \rightarrow 6C_2H_5I(l) + 2H_3PO_3(l)$$

and **diazonium compounds** can be reacted with potassium iodide to produce iodoarenes; for instance:

$$C_6H_5N_2^+(aq) + I^-(aq) \rightarrow C_6H_5I(l) + N_2(g).$$

◀ Substitution ▶

HALOGEN CARRIER

A halogen carrier is a **catalyst**, such as aluminium chloride or iron(III) chloride, which promotes the **electrophilic substitution** of **aromatic** compounds by chlorine and bromine. Halogen carriers are typically **Lewis acids**, which react with the halogen by accepting a **lone pair** of electrons from one of its atoms:

$$Cl-Cl: + AlCl_3 \rightarrow Cl^+ \ Cl-AlCl_3^-$$

Aluminium chloride is the catalyst used in **Friedel–Craft reactions**. Iron(III) chloride or bromide is the carrier commonly used in the halogenation of benzene and its derivatives.

◀ Electrophilic substitution ▶

HALOGENOALKANES

Halogenoalkanes are the compounds formed by substituting halogen **atoms** for hydrogen atoms in an **alkane**. These organic halogen compounds may be grouped into a variety of **homologous series** depending upon the number and type of halogen and upon the nature of the carbon skeleton. For example, one of the simplest homologous series is the 1-chloroalkanes with no branching in the carbon chain.

 NOMENCLATURE

In the systematic method of naming halogenoalkanes the root name is based on the longest continuous carbon chain and the **straight-chain alkane** with the same number of carbons. The root name is prefixed with the numbers (**di-, tri-, tetra-,** etc.) and names of the **halogens** (**fluoro-, chloro-, bromo-, iodo-**). The positions of the halogens are given the lowest possible numbers and the prefixes are attached in alphabetical order of halogen. For example, the systematic name of the widely used anaesthetic 'Halothane' is:

$$
\begin{array}{c}
\quad\;\; F \;\; Cl \\
\quad\;\; | \quad\; | \\
F-C-C-Br \\
\quad\;\; | \quad\; | \\
\quad\;\; F \;\; H
\end{array}
$$
2-bromo-2-chloro-1,1,1-trifluoroethane

Some simple compounds are still traditionally referred to as primary, secondary or tertiary halogenoalkanes; for instance:

$$
\begin{array}{ccc}
\quad H & \quad CH_3 & \quad CH_3 \\
\quad | & \quad | & \quad | \\
H-C-Cl & CH_3-C-Br & CH_3-C-I \\
\quad | & \quad | & \quad | \\
\quad H & \quad H & \quad CH_3
\end{array}
$$

traditional:	methyl chloride (primary)	isopropyl bromide (secondary)	tertiary butyl iodide (tertiary)
systematic:	chloromethane	2-bromopropane	2-iodo-2-methylpropane

PHYSICAL PROPERTIES

The **densities, melting points** and **boiling points** (see Fig. H.4) of the halogenoalkanes are higher than those of the corresponding alkanes and the values increase with increasing **atomic number** of the halogen and with increasing halogenation of the alkane.

Name	Formula	$T_b/°C$	Name	Formula	$T_b/°C$
chloromethane	CH_3Cl	−24	fluoroethane	C_2H_5F	−38
dichloromethane	CH_2Cl_2	40	chloroethane	C_2H_5Cl	12
trichloromethane	$CHCl_3$	62	bromoethane	C_2H_5Br	38
tetrachloromethane	CCl_4	77	iodoethane	C_2H_5I	72

Fig. H.4 Boiling temperatures of some halogenoalkanes

▶ *CHEMICAL PROPERTIES*

The halogenoalkanes undergo **nucleophilic substitution** and/or **elimination reactions** depending upon the strength and polarity of the carbon–halogen bond, the structure of the alkyl **group** and the nucleophile's strength as a **base** (proton acceptor). The carbon atom becomes more electrophilic and heterolysis of the C–Hal bond more likely as the halogen becomes more electronegative. Primary halogenoalkanes favour substitution reactions and tertiary halogenoalkanes favour elimination reactions, while strongly basic nucleophiles favour elimination reactions and weakly basic nucleophiles favour **substitution reactions**. Consequently, the reactions of the halogenoalkanes with alkali metal hydroxides tend to be **hydrolysis** in aqueous conditions and **dehydrohalogenation** in alcoholic conditions:

$$C_4H_9Br(l) + NaOH(aq) \xrightarrow[\text{reflux}]{\text{heat under}} C_4H_9OH(l) + NaBr(aq)$$

$$C_4H_9Br(l) + KOH(alc) \xrightarrow[\text{reflux}]{\text{heat under}} C_2H_5CH{=}CH_2(g) + H_2O(l) + NaBr(aq)$$

The reactivity of the halogenoalkanes increases as the strength of the carbon–halogen bond decreases (from C–F to C–I). Consequently, fluorocarbons and chlorofluorocarbons such as Poly(TetraFluoroEthene) (PTFE) and the FREONS are used when chemical inertness and thermal stability are needed. Chloro- and bromo-alkanes are used in organic syntheses because they are sufficiently reactive to be converted into **alcohols, ethers, esters, nitriles, amines,** and **Grignard reagents**. The iodoalkanes are relatively unstable and may decompose even when stored in brown bottles to keep out light.

HEAT OF REACTION

Heat of reaction is the term which refers loosely to the heat change of an **exothermic** or **endothermic chemical reaction**.

The heat change measured at constant pressure is called the **enthalpy change**, $\triangle H$. The heat change measured at constant volume is called the **internal energy change**, $\triangle U$. The two types of heat change are related by the expression $\triangle H = \triangle U + \triangle nRT$, where R is the **gas constant**, T is the absolute temperature and $\triangle n$ = no. moles gaseous products − no. moles gaseous reactants.

◀ Enthalpy change ▶

HESS'S LAW

The **standard molar enthalpy change** of a process is independent of the means or route by which that process takes place.

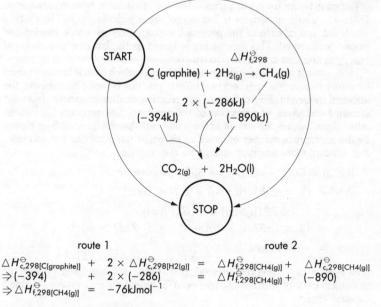

route 1   route 2

$\Delta H^{\ominus}_{c,298[C(graphite)]} + 2 \times \Delta H^{\ominus}_{c,298[H2(g)]} = \Delta H^{\ominus}_{f,298[CH4(g)]} + \Delta H^{\ominus}_{c,298[CH4(g)]}$

$\Rightarrow (-394) + 2 \times (-286) = \Delta H^{\ominus}_{f,298[CH4(g)]} + (-890)$

$\Rightarrow \Delta H^{\ominus}_{f,298[CH4(g)]} = -76 \text{kJmol}^{-1}$

Fig. H.5 Calculating enthalpy change using Hess's law

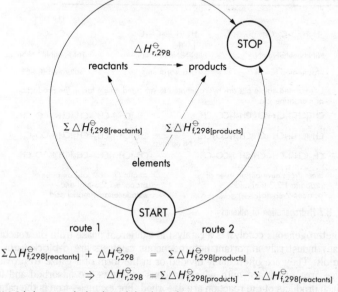

route 1   route 2

$\Sigma \Delta H^{\ominus}_{f,298[reactants]} + \Delta H^{\ominus}_{r,298} = \Sigma \Delta H^{\ominus}_{f,298[products]}$

$\Rightarrow \Delta H^{\ominus}_{r,298} = \Sigma \Delta H^{\ominus}_{f,298[products]} - \Sigma \Delta H^{\ominus}_{f,298[reactants]}$

Fig. H.6 Calculating standard molar enthalpy change for any reaction, using Hess's law

Germain Henri Hess put forward his law of constant heat summation in 1840. The above statement is just one of many expressions of Hess's law, which is a special case of the general law that *energy cannot be created and cannot be destroyed*. This general law is known as the law of conservation of energy or the first law of thermodynamics.

Hess's law is used to calculate energy changes which cannot be determined by direct measurement (see Fig. H.5). It can also be used to calculate the standard molar enthalpy change of any reaction, real or imaginary, from the standard enthalpies of formation of the reactants and products. Σ (Greek letter sigma) means the sum of all the molar enthalpy changes, each multiplied by the appropriate number of moles as shown by the equation. For example, the standard molar enthalpy change for the reaction

$$H_2S(g) + Cl_2(g) \rightarrow 2HCl(g) + S(s) \quad \text{is given by}$$

$$\begin{aligned}\Delta H_r^\theta &= \{2 \times \Delta H_f^\theta [HCL(s)] + \Delta H_f^\theta [CL_2(g)]\} \\ &\quad - \{\Delta H_f^\theta [H_2S(g)] + \Delta H_f^\theta [S(s)]\} \\ &= \{2 \times (-92.3) + 0)\} - \{1 \times (-20.6) + 0\} \\ &= -164 \text{ kJ mol}^{-1}\end{aligned}$$

The sign (+ or −) and the units (kJ mol^{-1}) should always be given with the numerical value of enthalpy change, even for an endothermic process having a positive sign.

HETEROGENEOUS CATALYST

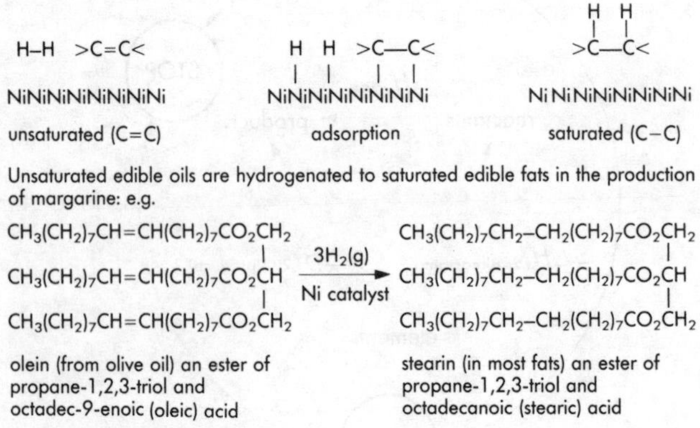
H–H >C=C<
NiNiNiNiNiNiNiNi
unsaturated (C=C)
H H >C—C<
NiNiNiNiNiNiNiNi
adsorption
H H
>C—C<
Ni NiNiNiNiNiNiNi
saturated (C−C)
Unsaturated edible oils are hydrogenated to saturated edible fats in the production of margarine: e.g.
CH3(CH2)7CH=CH(CH2)7CO2CH2
CH3(CH2)7CH=CH(CH2)7CO2CH
CH3(CH2)7CH=CH(CH2)7CO2CH2
3H2(g)
Ni catalyst
CH3(CH2)7CH2–CH2(CH2)7CO2CH2
CH3(CH2)7CH2–CH2(CH2)7CO2CH
CH3(CH2)7CH2–CH2(CH2)7CO2CH2
olein (from olive oil) an ester of propane-1,2,3-triol and octadec-9-enoic (oleic) acid
stearin (in most fats) an ester of propane-1,2,3-triol and octadecanoic (stearic) acid

Fig. H.7 Hydrogenation of alkenes

A heterogeneous **catalyst** is a catalyst in a different phase from the reactants. Many industrially important heterogeneous catalysts are **d-block transition metals**. They are often finely divided or in the form of a gauze to provide a large area of surface onto which gaseous reactants are adsorbed and from which products of the reaction are desorbed. For example, iron is the catalyst in the **Haber synthesis** of ammonia from nitrogen and hydrogen while chromium, cobalt, copper and nickel are catalysts for a range of reactions in the **petrochemical** industry. A platinum–rhodium gauze is used to catalyse the **oxidation** of ammonia to nitrogen oxide in the industrial production of nitric acid. These metal catalysts use their 3d and 4s subshells in the **adsorption** of the reactant molecules to lower **activation energies** by weakening bonds in the adsorbed molecules. The **hydrogenation** of an **alkene** using a nickel catalyst is shown in Fig. H.7.

◀ Catalyst ▶

HETEROLYTIC FISSION

Heterolytic fission is the breaking of a **single covalent bond** between two atoms so that one **atom** retains both of the shared **electrons**. It is most likely to occur with a **polar bond** between two atoms with significantly different **electronegativities**. For example, in the reactions of chloroalkanes, the carbon–chlorine bond is usually considered to break 'heterolytically' (into two unequal parts):

$$-C\overset{\frown}{-}Cl \quad \rightarrow \quad -C^+ \quad :Cl^-$$

chlorine atom departs with both electrons to form a stable anion

In heterolysis, the atom departing with the bonding **electron pair** leaves behind an electrophilic carbon atom vulnerable to attack by a **nucleophile**. The movement of the electron pair is shown in a **reaction mechanism** by a curly arrow.

◀ Electrophile, Nucleophile ▶

HEXAGONAL CLOSE-PACKING

Hexagonal close-packing (Fig. H.8) refers to the packing of identical spheres into an arrangement occupying the minimum space, such that the spheres in

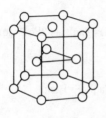
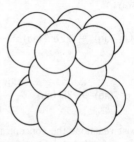

hexagonal close-packed (hcp)

Fig. H.8

the third layer are directly above the spheres in the first layer, giving the ABABAB pattern. The coordination number in hexagonal close packing is 12.

◀ Cubic close-packing ▶

HOFMANN DEGRADATION

Hofmann degradation is a reaction in which an amide is converted into an **amine** with a shorter carbon chain. In the Hofmann reaction, an amide is heated with bromine and aqueous alkali to form a primary amine and lose a carbon atom as CO_2. This reaction is one way of precisely shortening a carbon chain:

$$RCH_2CONH_2 + Br_2 + 2NaOH \rightarrow RCH_2NH_2 + CO_2 + 2NaBr + H_2O$$

The Hofmann degradation also refers to the formation of an **alkene** and a tertiary amine by heating a **quaternary ammonium** hydroxide. August Wilhelm Hofmann (1818−92) was a professor of chemistry at Berlin and at the Royal College of Chemistry in London.

HOMOGENEOUS CATALYST

A homogeneous **catalyst** is a catalyst in the same phase as the reactants. Homogeneous catalysts take part in a reaction, so an increase in their **concentration** will increase the speed of the **rate-determining step**.

The **oxidation** of iodide **anions** by peroxodisulphate(VI) anions is slow:

$$S_2O_8^{2-}(aq) + 2I^-(aq) \xrightarrow[\Rightarrow \text{ slow}]{\text{high } E_a} 2SO_4^{2-}(aq) + I_2(aq)$$

Iron(II) or iron(III) **cations** catalyse the reaction by providing an alternative pathway:

$$S_2O_8^{2-}(aq) + 2Fe^{2+}(aq) \xrightarrow[\Rightarrow \text{ fast}]{\text{low } E_a} 2SO_4^{2-}(aq) + 2Fe^{3+}(aq)$$

$$2Fe^{3+}(aq) + 2I^-(aq) \xrightarrow[\Rightarrow \text{ fast}]{\text{low } E_a} 2Fe^{2+}(aq) + I_2(aq)$$

In most reactions the concentration of catalyst remains constant with time, but in an **autocatalytic** reaction the concentration increases with time, because one of the products acts as a catalyst. A reaction is autocatalytic if it is catalysed by one of its products.

The iodination of propanone is autocatalytic because the reaction is **first-order** with respect to $H^+(aq)$, and aqueous hydrogen ions are produced in the reaction:

$$CH_3COCH_3(aq) + I_2(aq) \xrightarrow[\text{catalyst}]{\text{acid}} CH_3COCH_2I(aq) + H^+(aq) + I^-(aq)$$

$$\uparrow$$

product acts as catalyst

HOMOLOGOUS SERIES

A series of related organic compounds in which the **formula** of each member differs from the preceding member by CH_2.

Compounds are divided into classes (**hydrocarbons, alcohols, carboxylic acids** and so on) on the basis of their composition and properties. Each class is subdivided into various homologous series (**alkanes, primary monohydric alcohols, straight-chain fatty acids** and so on) on the basis of their composition and properties. Each homologous series is distinguished by its general formula.

Members of a homologous series always have the same functional group(s) and show both similarities and trends in their properties. The *similarity* in properties is explained by the same **functional group(s)** in each molecule. The *trend* is explained by the change in **intermolecular forces** and molecular complexity with the increasing molecular size. Two compounds with the same functional group need not be members of the same homologous series. For example, ethanol and cyclohexanol are not homologues.

The names of the homologous series of alkanes play a particularly important part in the **nomenclature** of many organic compounds.

HOMOLYTIC FISSION

Homolytic fission is the breaking of a **single covalent bond** between two **atoms** so that each atom retains only one of the shared **electrons**. Ultraviolet light initiates the chlorination and bromination of alkanes by causing homolytic fission of chlorine and bromine molecules. For example:

$$Br-Br \xrightarrow{\text{UV light}} 2Br\cdot$$

In organic reactions, $C-H$ and $C-C$ bonds are usually considered to break 'homolytically' (into two equal parts); for example:

$$-\overset{|}{\underset{|}{C}}-\overset{|}{\underset{|}{C}}- \rightarrow -\overset{|}{\underset{|}{C}}\cdot \quad \cdot\overset{|}{\underset{|}{C}}-$$ each carbon atom keeps one of the electrons from the bonding pair

Homolysis produces molecular fragments with unpaired electrons called **free radicals**. These fragments are very reactive and take part in **chain reactions**. Homolytic fission should not be confused with **heterolytic fission**.

◄ Radical chain mechanism ►

HYDRATE

A hydrate is a substance formed when water combines with a compound, usually without the water decomposing. Hydrates are commonly formed by ionic compounds such as salts. Sodium sulphate, for example, forms a mono-, hepta- and deca-hydrate: $Na_2SO_4.H_2O$, $Na_2SO_4.7H_2O$ and $Na_2SO_4.10H_2O$. These are now called sodium sulphate-1-water, sodium sulphate-7-water and sodium sulphate-10-water, respectively. The water in these hydrates is known as **water of crystallisation**. It can usually be removed by heating to produce the anhydrous salt.

The compound formed when water combines with trichloroethanal is commonly known as chloral hydrate:

$$
\begin{array}{ccc}
\text{Cl} & & \text{Cl OH} \\
| & \quad\;\;\; \text{O} & | \; | \\
\text{Cl}-\text{C}-\text{C} & + \text{H}_2\text{O} \rightarrow & \text{Cl}-\text{C}-\text{C}-\text{H} \\
| & \text{H} & | \; | \\
\text{Cl} & & \text{Cl OH}
\end{array}
$$

HYDRATION

Hydration is the combination of a substance with water. **Alkenes** can be hydrated to form **alcohols** – pure ethanol, for instance, is manufactured on the industrial scale by the hydration of ethene obtained from oil and natural gas:

$$
\text{CH}_2{=}\text{CH}_2 + \text{H}_2\text{O} \xrightarrow[\substack{\text{phosphoric acid catalyst} \\ \text{on a silica support}}]{60\text{–}70 \text{ atm at } 300°C} \text{CH}_3{-}\text{CH}_2\text{OH}
$$

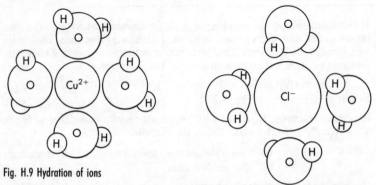

Fig. H.9 Hydration of ions

The process of hydration involves the attachment of **polar water molecules** to **ions** by **ion–dipole** electrostatic forces of attraction to form hydrated ions.

In aqueous solution, ions are surrounded by a sheath of oriented water molecules (see Fig. H.9). The hydration energy given out when an ionic substance dissolves in water is comparable to the **lattice energy** taken in when the crystalline solid breaks up. The small difference between the hydration energy and the lattice energy is the energy change for the dissolving of the ionic substance.

HYDRATION NUMBER

The hydration number is the number of **molecules** in the sheath of water surrounding and moving with an **ion** in **aqueous** solution. It is not an exact number. It depends upon the charge and size of the ion and upon the method used to determine the number. The approximate hydration numbers of Na^+, Mg^{2+} and Al^{3+} are 5, 15 and 30 respectively.

HYDRIDES

A hydride is a compound of an element and hydrogen. The **s-block metals** form *ionic hydrides* containing the H^- anion which react readily with water to form hydrogen; for instance:

$$NaH(s) + H_2O(l) \rightarrow NaOH(aq) + H_2(g)$$

Electrolysis of these molten hydrides produces hydrogen at the positive electrode. Some **transition metals** form interstitial hydrides wherein hydrogen atoms are held in the holes (interstices) between the metal atoms to give a hard metallic-like, non-stoichiometric solid. Except for the **noble gases**, most non-metallic elements form covalent hydrides, many of which have a simple molecular structure. For example, the hydrogen halides are simple **diatomic molecules** – see Fig. H.10.

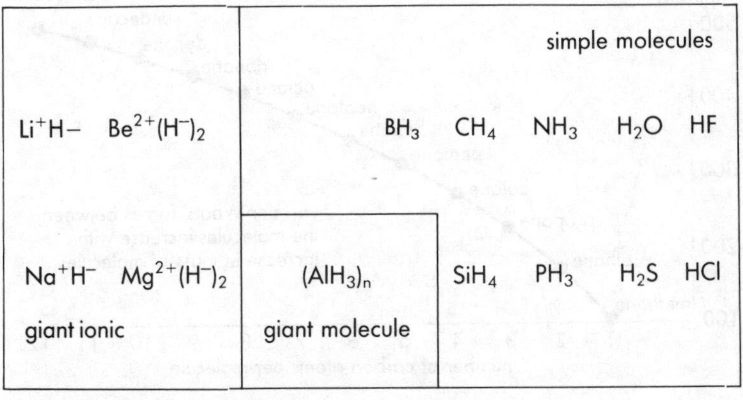

						simple molecules
Li^+H-	$Be^{2+}(H^-)_2$		BH_3	CH_4	NH_3	H_2O HF
Na^+H^-	$Mg^{2+}(H^-)_2$	$(AlH_3)_n$	SiH_4	PH_3	H_2S	HCl
giant ionic		giant molecule				

Fig. H.10 Trend in properties of hydrides

For patterns in covalent–ionic character of hydrides of the elements, see Fig. C.21.

HYDROCARBONS

Hydrocarbons are compounds of carbon and hydrogen only. They are mainly derived from coal, natural gas, petroleum and plants. They may be **saturated** or **unsaturated**, and their molecules may be in the form of chains or rings. Alkanes (C_nH_{2n+2}) are saturated chain compounds and **cycloalkanes** (C_nH_{2n}) are saturated ring compounds. **Alkenes** (C_nH_{2n}) and **alkynes** (C_nH_{2n-2}) are unsaturated chain compounds whilst **cycloalkenes** (C_nH_{2n-n}) are unsaturated ring compounds. The **aromatic** hydrocarbons (**arenes**) form a major class of ring compounds consisting of the benzene, napthalene and anthracene groups.

▶ PHYSICAL PROPERTIES

The physical state of a hydrocarbon depends upon its **molar mass** as well as the temperature and pressure. Under ordinary conditions the hydrocarbons of low molar mass are gaseous or liquid but those with molar mass above 130 g mol^{-1} are solid. Hydrocarbons do not mix with water because more energy is required to disrupt the **dipole–dipole** and **hydrogen–bonding** forces between the water molecules than can be provided by the forming of **van der Waals forces** between the hydrocarbon and water molecules in the mixture. Hydrocarbons float on water because the structural units $-CH_2^-$ and H_2O have similar volumes but the **relative atomic mass** of oxygen is greater than that of carbon. Hence hydrocarbons are less dense than water. The boiling temperature of the hydrocarbons generally decreases with increasing complexity of chain branching and increases with increasing molar mass as the

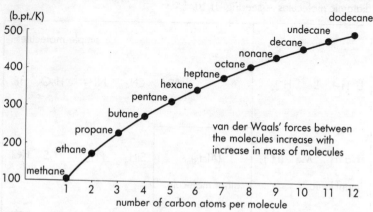

Fig. H.11 Boiling temperatures of straight-chain alkanes

van der Waals forces of attraction increase with the number of **atoms** in the molecule; see Fig. H.11.

▶ CHEMICAL PROPERTIES

Hydrocarbons burn. The reaction always produces heat and water. It is often explosive. If the supply of oxygen is inadequate, carbon monoxide and carbon may be produced as well as, or instead of, carbon dioxide. Complete combustion in oxygen under pressure in a **bomb calorimeter** always produces carbon dioxide and water:

$$C_xH_y + (x + y/4)O_2 \rightarrow xCO_2 + y/2H_2O$$

Explosions are caused by a **free-radical chain mechanism** producing an extremely fast **exothermic reaction** whose heat causes a sudden expansion in the volume of the gaseous combustion products. Heat is given out because less energy is required to break the carbon–carbon and carbon–hydrogen bonds in the C_xH_y molecules and the oxygen–oxygen bonds in the O_2 molecules than is given out when the carbon–oxygen bonds form to produce the CO_2 molecules and the hydrogen–oxygen bonds form to produce the H_2O molecules.

In general, as x in the formula C_xH_y increases, the hydrocarbons become more viscous and less volatile and, not surprisingly, more difficult to ignite and less likely to burn completely in air. Saturated hydrocarbons of high molar mass and unsaturated hydrocarbons burn with a yellow smoky flame.

Saturated hydrocarbons can be halogenated by free-radical substitution reactions and are generally less reactive than unsaturated aliphatic hydrocarbons, which undergo a wider range of **addition reactions**. The unsaturated aromatic hydrocarbons (arenes) show an even greater range of mainly **electrophilic substitution reactions**.

HYDROGENATION

Hydrogenation is the catalysed reaction of an **unsaturated** organic compound with hydrogen to form a **saturated** compound.

For the industrial production of **nylon-6**, the unsaturated benzene ring is reduced by hydrogen and nickel, as a **heterogenous catalyst**, to the saturated cyclohexane ring. This reaction releases about 150kJ less heat than would be expected from the hydrogenation of three moles of C=C double bonds. The

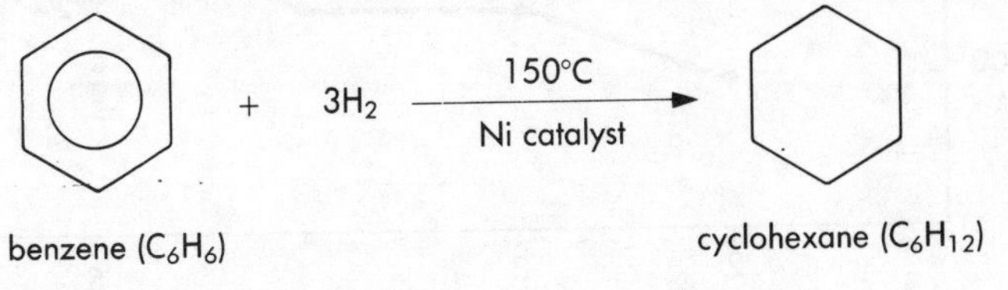

benzene (C_6H_6)   cyclohexane (C_6H_{12})

Fig. H.12 Hydrogenation of benzene

150kJ is the delocalisation enthalpy and a measure of the extra stability of the benzene ring. The production of margarine involves the hydrogenation of some unsaturated edible oils to produce saturated edible fats: see Fig. H.7.

The term hydrogenation also refers to the production of liquid hydrocarbons from coal by heating with hydrogen and catalysts at high pressures.

HYDROGEN BONDING

A hydrogen bond ($\cdots$) is a weak bond between a very electronegative atom (X = N, O or F) and a hydrogen atom bonded to a very electronegative atom (Y = N, O or F), thus $-X \cdots H-Y$.

The bond can be explained by saying that the highly electronegative atom Y polarises the H–Y bond so much that the nucleus of the hydrogen atom is exposed to attraction by a lone pair on another electronegative atom X. Hydrogen bonding is stronger than van der Waals forces but weaker than covalent bonding. It is a major factor in determining the structure and properties of water, hydrated salts, carbohydrates and proteins. The exceptionally high boiling points of ammonia, water and hydrogen fluoride (compared to the values for the other hydrides in each group) are explained by hydrogen bonding increasing the intermolecular forces of attraction (see Fig. H.13). Hydrogen bonding between the carboxylic acid groups accounts for the dimerisation of monoprotic carboxylic acids such as ethanoic acid and benzoic acid in non-ionising solvents such as methylbenzene – see Fig. H.14.

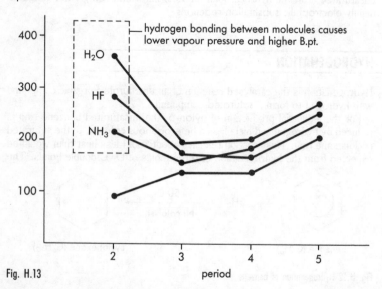

Fig. H.13   period

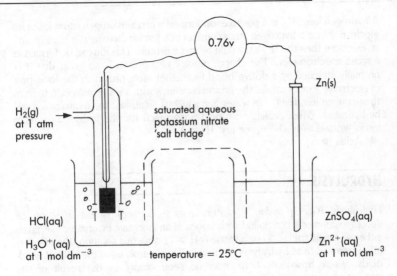

ethanoic acid dimer

benzoic acid dimer

Fig. H.14 Dimerisation caused by hydrogen bonding

HYDROGEN ELECTRODE

$$Pt[H_2(g)] \mid 2H_3O^+(aq) \parallel Zn^{2+}(aq) \mid Zn(s) \qquad E^\theta = -0.76v$$

Fig. H.15

A hydrogen electrode is an **electrochemical half-cell** consisting of pure hydrogen gas bubbling past a platinised platinum electrode dipping into **aqueous acid** (Fig. H.15). Platinum is used to make electrical contact with the hydrogen gas and aqueous **hydrogen ions** because the metal is too inert to act as a reductant. The metal surface is covered with platinum black – a finely divided coating of platinum deposited electrolytically. At this electrode the following reversible electron-transfer reaction occurs between the hydrogen gas and the aqueous hydrogen ions:

$$H_2(g) + 2H_2O(l) \rightleftharpoons 2H_3O^+(aq) + 2e^-$$

or $\qquad H_2(g) \rightleftharpoons 2H^+(aq) + 2e^-$

The standard hydrogen electrode is the half-cell consisting of pure hydrogen gas, at 25°C and 1 atm pressure, bubbling past a platinised platinum electrode dipping into a 1 mol dm^{-3} solution of H_3O^+(aq). The standard hydrogen electrode is the primary reference electrode for **standard electrode potentials**. By definition, the standard electrode potential for the hydrogen electrode is zero. The electrode is not very convenient to use, so for most practical purposes it is replaced by a calomel electrode or some other secondary-reference electrode.

◀ Electrode potential ▶

HYDROGEN ION

A hydrogen ion, H^+, is a positive **ion** formed when a hydrogen **atom** loses its **electron**. Since a hydrogen atom consists of a proton forming the nucleus and an electron, then the H^+ ion would be just a proton. This ion can be formed in a **mass spectrometer**. The charge density of a proton is so great that it is normally attached by a **dative bond** to another atom providing the **lone-pair** of electrons. For example, the proton combines with a water molecule to form the oxonium ion, H_3O^+. In water and aqueous solutions, the oxonium ion will be hydrated. When dealing with **acid-base** reactions, the hydrated oxonium ion is written as H_3O^+(aq) or just H^+(aq).

◀ Acids ▶

HYDROLYSIS

Hydrolysis (Greek *hydor* – water; *lysis* – loosen) is the reaction or decomposition of a compound with water. If an inorganic or organic salt gives either an acidic (**pH < 7**) or alkaline (**pH > 7**) **aqueous** solution, the effect is often referred to as hydrolysis of the salt. This hydrolysis of the salts of **weak acids**, **weak bases**, or both, may be seen simply as the result of the interaction with water of one or other or both of the **ions** in the salt. For example, aqueous ammonium ethanoate is neutral because the **cation** and **anion** are both 'hydrolysed':

$$NH_4^+(aq) + H_2O(l) \rightleftharpoons NH_3(aq) + H_3O^+(aq)$$

$$CH_3CO_2^-(aq) + H_2O(l) \rightleftharpoons CH_3CO_2H(aq) + OH^-(aq)$$

The tendency of inorganic chlorides to hydrolyse increases as the character of the chloride changes from ionic to covalent. For example, chlorides become more covalent i) from left to right across the periodic table:

NaCl(s) MgCl$_2$(s)	AlCl$_3$(s) SiCl$_4$(l) PCl$_3$(l) S$_2$Cl$_2$(l)
water-soluble ionic compounds	covalent compounds hydrolysed by water to produce acidic solutions of HCl(aq)

and ii) with increasing **oxidation number** of the element as, for example, in Group IV, where all the tetrachlorides except CCl$_4$ are hydrolysed to the dioxide and HCl;

for instance:

$$SiCl_4(l) + 2H_2O(l) \rightarrow SiO_2(s) + 4HCl(aq)$$

Tetrachloromethane does not hydrolyse because the C-atom cannot extend its electron shell beyond eight, so a strong C–Cl bond must break before a C–O bond can form.

Hydrolysis of organic halogen compounds often involves nucleophilic attack of H_2O molecules or OH^- ions upon the electrophilic carbon atom to which the halogen is attached. **Halogenoalkanes** may be hydrolysed to **alcohols** slowly by an S_N1 mechanism if they are tertiary halogenoalkanes. **Aromatic** halogen compounds resist hydrolysis if the halogen is attached directly to the arene ring. **Acyl chlorides** hydrolyse very readily and must be kept out of contact with damp air:

$$CH_3C\overset{O}{\underset{Cl(l)}{<}} + H_2O(l) \rightarrow CH_3C\overset{O}{\underset{OH(l)}{<}} + HCl(g)$$

ethanoyl chloride   ethanoic acid

The slow hydrolysis of **amides, esters, nitriles** and **proteins** is usually catalysed by acid or alkali and carried out by heating under reflux:

$$CH_3C\overset{O}{\underset{NH_2(s)}{<}} + H_2O(l) \xrightarrow[\text{catalyst}]{NaOH(aq)} CH_3C\overset{O}{\underset{OH(l)}{<}} + NH_3(g)$$
ethanamide

$$CH_3C\overset{O}{\underset{OC_2H_5(l)}{<}} + H_2O(l) \xrightarrow[\text{catalyst}]{H_2SO_4(aq)} CH_3C\overset{O}{\underset{OH(l)}{<}} + C_2H_5OH(l)$$
ethyl ethanoate   ethanol

$$CH_3CN(l) + 2H_2O(l) \xrightarrow[\text{catalyst}]{HCl(aq)} CH_3C\overset{O}{\underset{OH(l)}{<}} + NH_3(g)$$

Alkaline hydrolysis of, for example, ethanoic acid derivatives $CH_3-\overset{O}{\overset{\|}{C}}-X$, proceeds by **nucleophilic addition** of OH^- to the C=O followed by an **elimination** of X^-; see Fig. H.16.

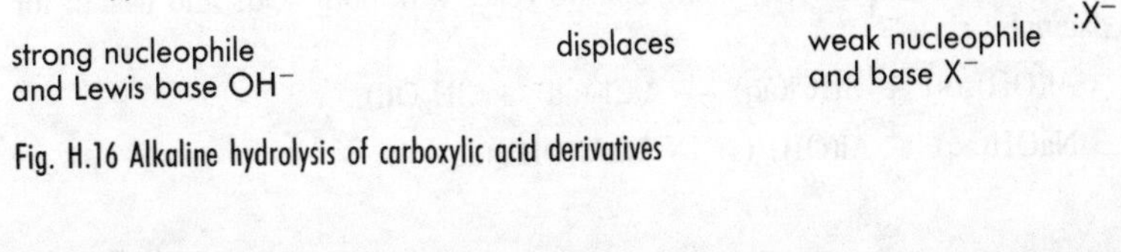

Fig. H.16 Alkaline hydrolysis of carboxylic acid derivatives

The weaker X^- is (as a nucleophile and base), the faster it is eliminated and the faster the hydrolysis takes place. In general the rate of hydrolysis of carboxylic acid derivatives increases from left to right as follows:

$$\text{slowest } \underset{\underset{O}{\|}}{RC}-NH_2 \rightarrow \underset{\underset{O}{\|}}{RC}-OR' \rightarrow \underset{\underset{O}{\|}}{RC}-O-CR' \rightarrow \underset{\underset{O}{\|}}{RC}-Cl \text{ fastest}$$

$\qquad$ amide $\qquad\qquad$ ester $\qquad\qquad$ anhydride $\qquad\qquad$ acyl chloride

HYDROPHILIC

Hydrophilic means 'water-loving' or having an affinity for water. In a surface active agent such as sodium stearate, the ionic carboxylate group $-CO_2^-$ is the *hydrophilic head*. A **hydrophilic sol** is a stable colloidal system consisting of a hydrophilic disperse phase (such as albumin or starch) in water as the dispersion medium. Amino ($-NH_2$), carboxyl ($-CO_2H$) and hydroxyl ($-OH$) groups are hydrophilic because they can form **hydrogen bonds** with water molecules.

◀ Detergents, Soaps ▶

HYDROPHOBIC

Hydrophobic means 'water-hating' or having no affinity for water. **Hydrocarbons** are hydrophobic. They do not mix with water since the energy required to break the **hydrogen-bonding** and **dipole—dipole attractions** between the water molecules cannot be provided by the weak **van der Waals forces** which are the only ones possible between the hydrocarbon and water. In a surfactant such as the octadecanoate anion, the long hydrocarbon chain is the *hydrophobic tail* which remains inside the micelle and does not mix with the water.

HYDROXIDES

Hydroxides are compounds of metals containing the hydroxide ion, OH^-, or the hydroxyl group, $-OH$.

Metal hydroxides are usually basic and sparingly soluble in water. The hydroxides of the **s-block metals** are strongly basic and those of Group I, the alkali metals, are very soluble in water. The solubility in water of the hydroxides of the Group II metals increases with increasing **atomic number** of the metal from the very sparingly soluble $Mg(OH)_2$ to the quite soluble $Ba(OH)_2$. The hydroxides of metalloids and metals such as aluminium, zinc, tin and lead are amphoteric, being able to react with both acids and bases; for example:

$$Al(OH)_3(s) + 3HCl(aq) \rightarrow AlCl_3(aq) + 3H_2O(l)$$

$$NaOH(aq) + Al(OH)_3(s) \rightarrow NaAl(OH)_4(aq)$$

The hydroxides of the non-metals are acidic and usually called oxoacids. For example, the formula of sulphuric acid is usually written as H_2SO_4 but it could be written as $(HO)_2SO_2$ to emphasise the existence of the two **hydroxyl groups** in the structure.

HYDROXYL GROUP

The hydroxyl group is the $-OH$ **functional group** present in water and **alcohols** and is responsible for their characteristic properties.
◀ Alcohols, Carboxylic acids ▶

IDEAL GAS EQUATION

The ideal gas equation states that:

$$pV = nRT$$

where p is the pressure, V the volume, T the temperture in kelvin, n the amount of gas in moles and R is a fundamental constant known as the **gas constant**.

The ideal gas equation incorporates **Boyles law, Charles law** and **Avogadro's principle** into one ideal gas law. The equation may be derived from the **kinetic molecular theory** of gases.

◄ Gas laws ►

IDEAL SOLUTION

An ideal solution is a solution of two volatile liquids whose total **vapour pressure** (at constant T) will show linear variation with composition.

Ideal solutions obey **Raoult's law** and are formed by liquids whose molecules have very similar structures and **intermolecular forces**: for instance, benzene and methylbenzene (two similar **arenes**), 2– and 3–methylpentane (two chain isomers of a **saturated hydrocarbon**) or pentanol or hexanol (two successive members of a **homologous series**). When liquids mix to form an ideal solution, there is little or no **enthalpy change** because the energy consumed in overcoming the forces holding the molecules together in the separate pure liquids is provided by the same forces re-forming between the very similar molecules in the solution; for example:

benzene molecules	+	methylbenzene molecules	mix →	solution of benzene and methylbenzene molecules
	energy given out when van der Waals forces are set up			energy taken in when van der Waals forces are broken down

Non-ideal solutions disobey Rauolt's law by showing positive (or negative) deviations from the predicted total vapour pressure, and they have positive (or negative) enthalpies of mixing.

◄ Azeotropes, Raoult's law, Van der Waals forces ►

INDICATORS

Acid–base indicators are substances whose colour changes with pH. They are usually brightly coloured organic compounds whose **molecules** can lose or gain **protons**. The colour change accompanying the proton loss (or gain) is connected with a change in molecular structure. An acid–base indicator may be regarded as a **weak acid** (HIn) whose **aqueous** undissociated acid molecules have a different colour from its aqueous **conjugate base** (In⁻); see Fig. I.1.

NAME	WEAK ACID FORM	pK_a	CONJUGATE BASE FORM
methyl orange	red	3.5	yellow
methyl red	red	5.1	yellow
bromothymol blue	yellow	7.0	blue
phenolphthalein	colourless	9.3	violet

Fig. I.1 Colour changes of acid–base indicators

Indicator solutions are only very **dilute**, and they are diluted even further, because only two or three drops of indicator solution are used in a **titration**. Consequently the **buffer** capacity of indicators is too small for them to act as buffers.

▶ MID-POINT COLOUR AND pH RANGE

The colour of an aqueous acid–base indicator is governed by the ratio of the concentrations of its weak acid (HIn) and conjugate base (In⁻).

Even in extremely dilute solutions, the weak acid (HIn) and conjugate base (In⁻) will be in equilibrium:

$$\frac{[H_3O^+(aq)][In^-(aq)]}{[HIn(aq)]} = K_{in} = K_a \text{ for indicators seen as weak acids.}$$

If HIn(aq) makes a solution yellow and In⁻(aq) makes it blue, then the solution will appear green if [HIn(aq)] = [In⁻(aq)]. Under these circumstances, [H_3O^+(aq)] = K_{in} and the colour is the mid-point colour of the indicator.

An acid–base indicator shows its mid-point colour when pH = pK_{in}. As a rule, our eyes will not detect one of the colours if it contributes less than 10 per cent to the overall colour. Consequently, an acid-base indicator changes colour over a range of about 2 pH units, one unit either side of the pH at its mid-point colour.

▶ SELECTING AN INDICATOR

For an acid-base indicator to be suitable for a titration, its pK_{in} should be equal to the pH at the **end-point** where pH changes most sharply.

Bromothymol blue or phenolphthalein (but not methyl orange) may be used as indicator when titrating 0.1 mol dm⁻³ aqueous propanoic acid with 0.1 mol dm⁻³ NaOH(aq) from a burette (see Fig. A.5). There is no indicator suitable for the titration of a weak acid with a weak alkali (or vice versa) because the pH does not change suddenly at the end–point, so no indicator will give a sharp change in colour.

▶ OTHER TYPES OF INDICATOR

In the widest sense, an indicator is a substance whose colour changes sharply in response to a change in concentration of a reactant (or product) involved in a stoichiometric reaction. Starch can be used an an indicator for iodine–thiosulphate titrations because it forms a blue-black complex with aqueous iodine molecules but is colourless in aqueous iodide ions. Ferroin, complex of iron ions and 1,10-phenanthroline, is used as an indicator for various redox titrations because the iron(II) form of the complex is intensely coloured but the iron(III) complex is colourless in aqueous solution. Murexide (a dyestuff) is used as an indicator for titrating aqueous calcium ions with EDTA because the colour of the calcium-murexide complex differs from the colour of the calcium-EDTA complex. Aqueous chromate(VI) ions are used as an adsorption indicator in titrating aqueous chloride ions with aqueous silver ions because silver chromate(VI) forms as a red solid on the surface of the white silver chloride precipitate.

◀ Titration, Titration curves, Volumetric analysis ▶

INFRA-RED ABSORPTION SPECTROSCOPY

Infra-red absorption spectroscopy is an analytical technique for determining the molecular structure of a substance by measuring the wavelengths of infra-red radiation absorbed by the substance.

Infra-red radiation emitted by a heated metal oxide rod is scanned, selected and focused by a system of mirrors and a prism or reflection grating, through the sample cell on to a thermal, photoelectric or photoconductive dectector. The sample cell (and prism if used) is made from an alkali metal halide such as sodium chloride or potassium bromide because glass absorbs infra-red radiation. The sample may be a gas, a thin (0.1mm) film of liquid pressed between two flat salt windows or a solid ground up with a liquid hydrocarbon and pressed between two flat salt windows. The output from the detector is automatically plotted, usually as percentage transmission, against the wavenumber of radiation transmitted (see Fig. I.2).

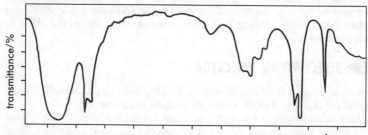

Fig. I.2 Infra-red absorption spectrum of ethanol   wave number/cm^{-1}

When the frequency of the infra-red radiation matches the bending or stretching frequencies of **polar bonds** within the molecules of the sample, absorption of the radiation occurs. For example, the stretching of the $C=O$ bond in **aldehydes** and **ketones** produces characteristic infra-red absorptions at wavenumbers between 1700 and 1740 cm^{-1}. $N-H$, $C-O$, $O-H$, $C=C$, $C\equiv C$ and $C\equiv N$ bonds also produce characteristic absorptions which allow their presence in a structure to be recognised. Even if the IR absorption spectrum of a substance cannot be used to elucidate its structure completely, the spectrum is unique and can always be used as a 'fingerprint' to distinguish one substance from another.

◀ Molecular spectroscopy ▶

INITIATION

The initiation stage in a **radical chain reaction** is the **homolytic fission** of **covalent bonds** to form **free radicals**.

Reactions involving gaseous chlorine or bromine are often initiated by ultraviolet light. The halogen molecules absorb the energy of the UV light and split to form Cl· or Br· radicals. Organic peroxides are used as initiators in **polymerisation reactions**. Their molecules absorb thermal energy by collision with other molecules and split to form free radicals; for example:

$$C_6H_5CO.O-O.COC_6H_5 \rightarrow 2C_6H_5COO \cdot \rightarrow 2C_6H_5 \cdot + 2CO_2.$$

INNER-SHELL SHIELDING

Inner-shell shielding (or screening) is the repulsion of complete inner shells of **electrons** upon the outer-shell **valence electrons** and the consequent reduction in the electrostatic force of attraction of the **nucleus** upon the outer electrons. The **valence electrons** in the outer shell of an atom are shielded more effectively by the electrons in the inner shells than by other valence electrons in the same outer shell. This explains on the one hand the large decrease in **atomic radius** and large increase in **first ionisation energy** from sodium to chlorine across the periodic table and on the other hand the relatively small changes in atomic radius and first ionisation energy from titanium to copper across the first transition series.

◀ Transition metals ▶

INORGANIC CHEMISTRY

Inorganic chemistry is the study of the elements and their compounds, apart from those compounds of carbon classified as organic compounds.

INTERMEDIATE BOND

An intermediate bond is one which is neither perfectly covalent nor completely ionic. Two non-metallic atoms having different **electronegativities** will tend to share electrons unequally and form a **polar covalent bond** which can be said to have some ionic character. The **polarising power** of a small, highly charged cation will tend to distort a large, negatively charged, highly polarisable anion and cause the **ionic bonding** to have some covalent character.

The factors governing the tendency for elements to form intermediate bonds are summarised by **Fajan's rules**.

INTERMOLECULAR FORCES

Intermolecular forces is a collective term for the **van der Waals forces**, **dipole—dipole interactions** and **hydrogen-bonding** that can arise between molecules.

INTERNAL ENERGY CHANGE

The internal energy change, $\triangle U$, of a process is the heat change for that process measured at constant volume. The **bomb calorimeter** measures the heat of combustion of substances at constant volume. This internal energy change is used to calculate the enthalpy of combustion. $\triangle U$ is related to $\triangle H$, the enthalpy change, by the expression:

$\triangle U = \triangle H - \triangle nRt$ where R is the gas constant, T is the absolute temperature and $\triangle n =$ no. moles gaseous products $-$ no. moles gaseous reactants.

◀ Enthalpy change ▶

ION

An ion is an **atom** or **group** of atoms that has gained or lost one or more electrons. The word ion (from a Greek verb meaning 'to go') was introduced by Michael Faraday in 1834 to describe the moving charged particles responsible for the electric current in an **aqueous** or **molten electrolyte** during **electrolysis**.

◀ Anion, Cation ▶

ION–DIPOLE INTERACTION

An ion–dipole interaction is the attraction between an **ion** and a **polar molecule**. The **hydration** of ions is the result of a **cation** attracting the oxygen end of a water molecule and an **anion** attracting the hydrogen end.

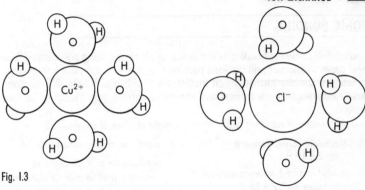

Fig. I.3

ION EXCHANGE

Ion exchange is the exchange of **ions** between an insoluble solid ion-exchanger and, usually, an **aqueous** solution in contact with it.

Ion-exchangers are anionic or cationic and may be natural or synthetic. Clay minerals and humus in the soil act as cationic exchangers for the **s-block metal** ions and H^+ ions. Natural and synthetic zeolites are inorganic **polymers** of hydrated aluminosilicates which exchange cations. The most versatile group of ion-exchangers are the synthetic organic cross–linked polymer resins. The anionic exchange resins contain positively charged **groups** such as $-NH_3^+$ to which **anions** are attached. The cationic exchange resins contain negatively charged groups such as $-CO^{2-}$ and $-SO^{3-}$ to which **cations** are attached.

In general, ion-exchange is performed simply by passing the aqueous solution of ions through a bed of the granular, porous ion-exchanger. In water softening, for example, the hard water is passed through a bed of a cationic exchanger which replaces the calcium and magnesium ions in the water with sodium ions. The reversible exchange may be represented as follows:

$$
\begin{array}{cccc}
Ca^{2+} & Ca^{2+} & & \\
Na^+ \quad Na^+ \quad Na^+ \quad Na^+ & & \rightleftharpoons & \\
X^- \quad X^- \quad X^- \quad X^- & & & \\
\end{array}
\qquad
\begin{array}{cccc}
Na^+ \quad Na^+ \quad Na^+ \quad Na^+ \\
Ca^{2+} \qquad Ca^{2+} \\
X^- \quad X^- \quad X^- \quad X^- \\
\end{array}
$$

When the ion-exchanger becomes depleted of its exchangeable ions and saturated with the exchanged ions, it must be regenerated by passing a concentrated solution of its exchangeable ions through the bed to reverse the ion-exchange process. Cationic water softeners are usually regenerated with a concentrated salt solution.

Ion-exchange has a wide variety of uses, including the separation of **isotopes**, the removal of iron from wine and calcium from milk and the recovery of uranium from acid solutions, valuable metals from wastes and chromates from plating baths. If hydrogen ions are exchanged for metal cations, the resulting acidic solution leaving the ion-exchanger can be titrated with alkali and the result used to determine the concentration of the metal cations in the original solution.

IONIC BONDING

Ionic (electrovalent) bonding is the result of **electrons** being transferred from one **atom** to another and the **ions** packing together into a **crystal lattice**. Ionic bonding is usually formed when metals react with non-metals. Metal atoms lose electrons to form cations and non-metals atoms gain electrons to form anions.

atomic number of rubidium is 37	atomic number of oxygen is 8
⇒ electronic configuration is Rb 2.8.18.8.1	⇒ electronic configuration is O 2.6
⇒ predict loss of 1 valence electron to leave Rb⁺ 2.8.18.8	⇒ predict gain of 2 electrons to valence shell to give O^{2-} 2.8
	⇒ predict $(Rb^+)_2O^{2-}$ as the formula

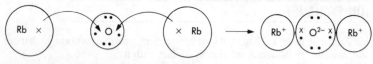

Fig. I.4 Predicting the formula of rubidium oxide

The formulae of simple binary ionic compounds of **s-block** (but not **d-block**) metals can be predicted from the **octet rule** and the ground state **electronic configurations** of the metal and non-metal atoms; see Fig. I.4. In general, ionic compounds form brittle, easily cleaved crystals with high melting points, high molar enthalpies of melting, poor electrical and poor thermal conductivity. When **aqueous** or molten, ionic compounds undergo **electrolysis**

The **crystal structures** and properties of caesium chloride, sodium chloride and calcium fluoride are typical of ionically bonded substances. The interpenetrating (or double) simple cubic structure of caesium chloride should not be confused with the **body-centred cubic structure** of the **alkali metals**. Caesium chloride is NOT a body-centred cube because the ion at the centre of the cube is different from the ions at the corners. In the drawings of **unit cells**, the lines between the ions merely serve to give perspective; they do NOT represent bonds (see Fig. I.5). The dimensions of the unit cell, and the arrangements of ions in it, may be determined by **X-ray** (diffraction) **crystallography**.

The heat of formation of an ionic compound may be seen mainly as the balance between the energy consumed in forming the gaseous ions and the energy released in forming the lattice. The **lattice energy** of an ionic compound depends upon the type of crystal structure and the charges and radii of the ions. An experimental value can be determined from a **Born–Haber cycle**. A completely theoretical value can be calculated on the assumption that the transfer of electrons from metal atoms to non-metal atoms is complete and the **cations** and **anions** are perfectly spherical. Comparison of these experimental and theoretical lattice energies indicates that alkali metal halides form ionic crystals with little or no covalent character but silver halides form crystals with considerable covalent character.

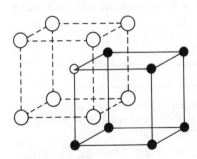 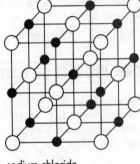

caesium chloride
double simple cubic

● = Cs^+ ○ = Cl^-

sodium chloride
face-centred cubic

● = Na^+ ○ = Cl^-

Fig. 1.5 Ionic crystal structures

The bonding between two elements is likely to be ionic when the difference between their **electronegativities** is large. **Fajan's rules** deal with the effect on the ionic bonding of the **polarising power** of the cation and the **polarisability** of the anion in terms of their charge and size.

IONIC CRYSTAL

An ionic crystal is a crystal with a lattice of separate **anions** and **cations**. Ionic crystals are formed by the alkali metal halides and by other metal salts. **Polarisation** of anions by strongly polarising cations often gives an otherwise ionic crystal some covalent character. The discrepancies between the experimental and theoretical lattice energies of the silver halides indicate that these compounds are not completely ionic.

◀ Crystal structure ▶

IONIC EQUATION

An ionic equation represents a reaction involving **ions** but does not show the **spectator ions** which are not taking part in the reaction. Ionic equations are most useful for **redox reactions** of **aqueous** ions. For example, the ordinary equation for the **oxidation** of aqueous sodium ethanedioate by potassium manganate(VII) in aqueous sulphuric acid is:

$$2KMnO_4(aq) + 8H_2SO_4(aq) + 5Na_2C_2O_4(aq) \longrightarrow$$
$$K_2SO_4(aq) + 2MnSO_4(aq) + 5Na_2SO_4(aq) + 8H_2O(l) + 10CO_2(g)$$

The ionic equation is:

$$2Mn_4^-(aq) + 16H^+(aq) + 5C_2O_4^{2-}(aq) \rightarrow 2Mn^{2+}(aq) + 8H_2O(l) + 10CO_2(g)$$

The ionic equation for the reactions of **strong aqueous acids** with **strong aqueous bases** is:

$$H_3O^+(aq) + OH^-(aq) \rightarrow 2H_2O(l)$$

IONIC PRODUCT OF WATER

The ionic product of water (K_w) is the constant value of the product of the concentrations ($[H_3O^+(aq)][OH^-(aq)]$) of the hydrogen and hydroxide ions in water and aqueous solutions at a given temperature:

$$K_w = 1 \times 10^{-14} \text{ mol}^2 \text{ dm}^{-6} \text{ at } 25°C \text{ and } 1 \times 10^{-13} \text{ mol}^2 \text{ dm}^{-6} \text{ at } 60°C.$$

The constant product is explained by the application of the **law of chemical equilibrium** to the following **reversible reaction**:

$$H_2O(l) + H_2O(l) \rightleftharpoons H_3O^+(aq) + OH^-(aq)$$

According to the law of chemical equilibrium:

$$\frac{[H_3O^+(aq)][OH^-(aq)]}{[H_2O(l)][H_2O(l)]} = K_c \text{ (at constant temperature)}$$

but the ionisation of the water is so slight that the **concentration** of the water molecules is regarded as constant. Hence

$$[H_3O^+(aq)][OH^-(aq)] = K_w$$

In pure water the concentration of $H_3O^+(aq)$ must be equal to the concentration of $OH^-(aq)$. So:

$$[H_3O^+(aq)] = [OH^-(aq)] = \sqrt{K_w} = 10^{-7} \text{ mol dm}^{-3} \text{ in pure water at } 25°C.$$

For convenience, pK_w is defined as $-\log_{10}(K_w/\text{mol}^2\text{dm}^{-6})$. The letter p (from the German *potenz* – power) is always in lower case, even where it is the first letter in a sentence. $pK_w = 14$ at 25°C.

◀ pH ▶

IONIC RADIUS

An ionic radius is a value for the radius of an ion treated as a sphere. Values for ionic radii are calculated from data and electron-density maps derived from **X-ray diffraction** analysis of **ionic crystals**. The distance between the **nuclei** of two neighbouring ions is very accurately determined for a variety of ions in a variety of crystals. From these distances, which are the sums of the radii of the two neighbouring ions, and a value for the radius of an ion chosen as a reference, a table of ionic radii is compiled. The ionic radius is not a precise quantity, because the size and shape of an ion depend upon such factors as the nature, number and arrangements of the ions surrounding it. The recorded values for ionic radius depend upon the method of measurement and the value of the ionic radius chosen for reference.

In spite of the uncertainties in their values, ionic radii of elements show some important patterns. The cationic radius is always smaller than the atomic radius of the metal forming the **cation**. The anionic radius is always larger than the atomic radius of the non-metal forming the **anion**. In any group of the periodic table the ionic radius increases with increasing atomic number down the group. As the charge on the cation increases from Li^+ to Be^{2+} to Al^{3+}, the ionic radius decreases. The same trend is shown for the sodium, magnesium and aluminium cations.

◀ Atomic radius ▶

IONISATION ENERGY

Ionisation energy is the minimum energy needed to remove a mole of **electrons** from a **mole** of gaseous ions. The energy required to remove one mole of electrons from one mole of gaseous atoms, $X(g)$, of an element is called the *molar first ionisation energy*. It is the energy (E_{m1}) for the process $X(g) \rightarrow X^+(g) + e^-$.

The energy required to remove one mole of electrons from one mole of the gaseous ions, $X^+(g)$, is called the *molar second ionisation energy* of the element. It is the energy (E_{m2}) for the process $X^+(g) \rightarrow X^{2+}(g) + e^+$.

In general, (E_{mj}) represents the *molar jth ionisation energy* of an element, and is the energy of the process $X^{(j-1)+}(g) \rightarrow X^{j+}(g) + e^-$. Ionisation energy values have been obtained by analysis of optical spectra and are listed in data books.

The pattern of successive ionisation energies indicates the arrangement of electrons into levels of energy. This arrangement is represented by the **electronic configuration** of the element (see Fig. I.6). The lowest ionisation energies correspond to the **valence electrons** which are involved in bonding. The higher ionisation energies correspond to the **inner shell electrons** which are not usually involved in bonding and which partially shield the valence electrons (by repulsion between similar charges) from the attraction by the nucleus.

logarithm of successive
ionisation energies (E_{mj})

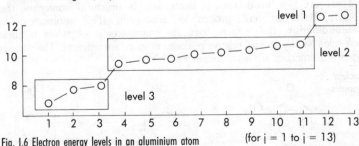

Fig. I.6 Electron energy levels in an aluminium atom (for j = 1 to j = 13)

The periodic pattern of molar first ionisation energies against **atomic number** indicates that the principal sets of electron energies may be arranged into subsets – see Fig. I.7.

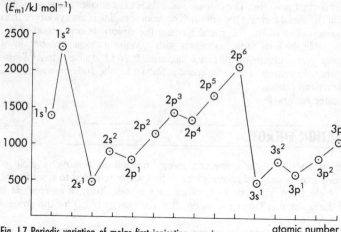

Fig. I.7 Periodic variation of molar first ionisation energies   atomic number

◀ Electronic configuration ▶

ISOELECTRONIC

Isoelectronic means 'having the same number of electrons'. Molecules and ions that are isoelectronic have very similar shapes. For example, :N≡N: and :C≡O: are isoelectronic molecules, and the tetrahedral ammonium cation NH_4^+ is isoelectronic with the tetrahedral methane molecule CH_4.

ISOMERISM

Isomerism is the existence of two or more compounds having the same molecular formula but different physical and/or chemical properties.

There are two broad types of isomerism. In structural isomerism, the isomers have different groups of atoms in their molecules. In stereoisomerism, the isomers have the same groups of atoms in their molecules but the groups' relative positions in space are different. The various types of isomerism are illustrated in Fig. I.8.

nuclear
isomers

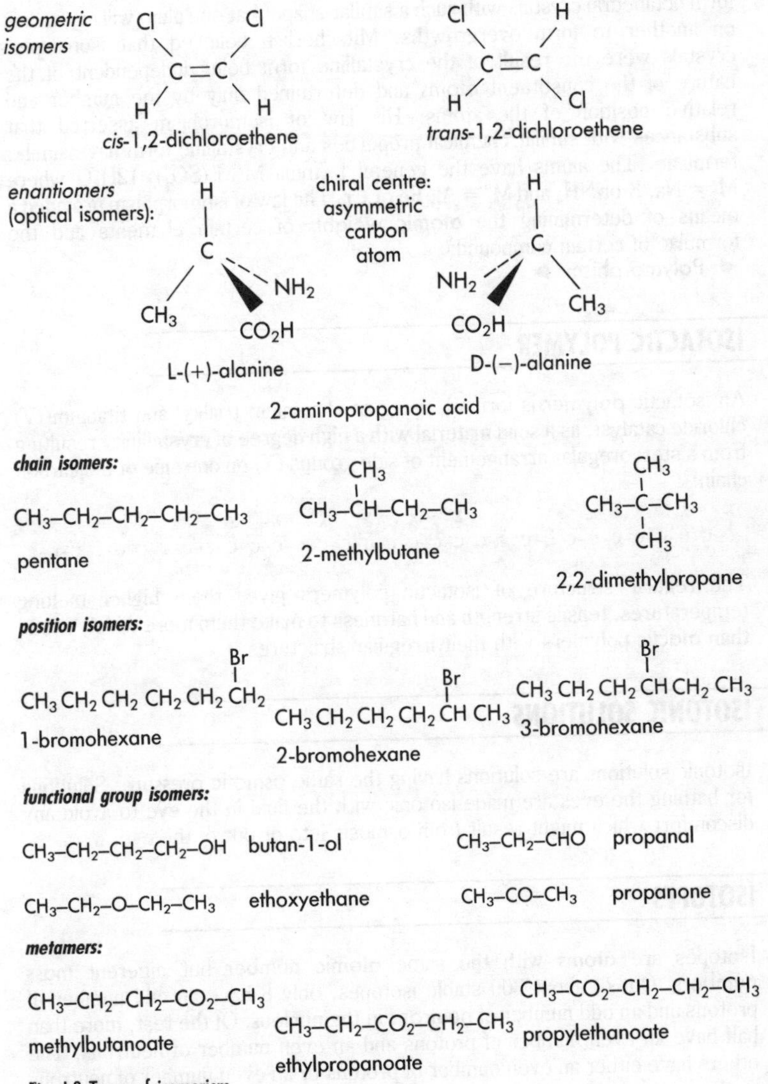

geometric isomers

cis-1,2-dichloroethene

trans-1,2-dichloroethene

enantiomers (optical isomers):

chiral centre: asymmetric carbon atom

L-(+)-alanine

D-(−)-alanine

2-aminopropanoic acid

chain isomers:

$CH_3–CH_2–CH_2–CH_2–CH_3$

pentane

$CH_3–CH–CH_2–CH_3$ (with CH_3 branch)

2-methylbutane

$CH_3–C–CH_3$ (with two CH_3 branches)

2,2-dimethylpropane

position isomers:

$CH_3 CH_2 CH_2 CH_2 CH_2 CH_2$ (Br)

1-bromohexane

$CH_3 CH_2 CH_2 CH_2 CH CH_3$ (Br)

2-bromohexane

$CH_3 CH_2 CH_2 CH CH_2 CH_3$ (Br)

3-bromohexane

functional group isomers:

$CH_3–CH_2–CH_2–CH_2–OH$ butan-1-ol

$CH_3–CH_2–CHO$ propanal

$CH_3–CH_2–O–CH_2–CH_3$ ethoxyethane

$CH_3–CO–CH_3$ propanone

metamers:

$CH_3–CH_2–CH_2–CO_2–CH_3$

methylbutanoate

$CH_3–CH_2–CO_2–CH_2–CH_3$

ethylpropanoate

$CH_3–CO_2–CH_2–CH_2–CH_3$

propylethanoate

Fig. 1.8 Types of isomerism

ISOMORPHISM

Isomorphism (*Greek iso* – same; *morphe* – form) is the ability of some different substances to form **crystals** with very similar external shape and internal structure. This feature was first reported in 1819 by Eilhard Mitscherlich, a professor of chemistry in Berlin. The double salts called alums form octahedral crystals with such a similar shape that one alum will crystallise on another to form overgrowths. Mitscherlich believed that isomorphic crystals were the result of the crystalline form being independent of the nature of the constituent atoms and determined only by the number and relative position of the atoms. His law of isomorphism asserted that substances with similar chemical properties and crystalline form have similar formulae. The alums have the general formula $M'M''(SO_4)_2 \cdot 12H_2O$ where $M' = $ Na, K or NH_4 and $M'' = $ Al, Fe or Cr. The law of isomorphism provided a means of determining the atomic weights of certain elements and the formulae of certain compounds.

◄ Polymorphism ►

ISOTACTIC POLYMER

An isotactic polymer is formed, using an aluminium trialkyl and titanium(IV) chloride catalyst, as a solid material with a high degree of crystallinity resulting from a stereoregular arrangement of side groups (X) on one side of the carbon chain:

$$-\overset{X}{\underset{|}{C}}-\overset{X}{\underset{|}{C}}-\overset{X}{\underset{|}{C}}-\overset{X}{\underset{|}{C}}-\overset{X}{\underset{|}{C}}-\overset{X}{\underset{|}{C}}-\overset{X}{\underset{|}{C}}-\overset{X}{\underset{|}{C}}-\overset{X}{\underset{|}{C}}-\overset{X}{\underset{|}{C}}-\overset{X}{\underset{|}{C}}-\overset{X}{\underset{|}{C}}-\overset{X}{\underset{|}{C}}-\overset{X}{\underset{|}{C}}-\overset{X}{\underset{|}{C}}-\overset{X}{\underset{|}{C}}-\overset{X}{\underset{|}{C}}-$$

The regular structure of isotactic polymers gives them higher melting temperatures, tensile strength and hardness to make them more widely useful than atactic polymers with their irregular structures.

ISOTONIC SOLUTIONS

Isotonic solutions are solutions having the same osmotic pressure. Solutions for bathing the eyes are made isotonic with the fluid in the eye to avoid any discomfort which might result from osmosis into or out of the eye.

ISOTOPES

Isotopes are atoms with the same atomic number but different mass numbers. Out of about 300 stable isotopes, only 8 have an odd number of protons and an odd number of neutrons in the nucleus. Of the rest, more than half have an even number of protons and an even number of neutrons. The others have either an even number of protons or an even number of neutrons in the nucleus. For atomic numbers up to 20, the ratio of the number of neutrons to the number of protons in stable isotopes is 1:1. As the atomic number increases, the stable-neutron-to-proton ratio increases to about 1.6:1. The natural isotopic abundances of elements have been very accurately determined with a mass spectrometer.

Some isotopes are unstable, and their nuclei spontaneously disintegrate, radiating energy in the process. These radioactive isotopes give out either alpha (α) or beta(β) particles (but never both) when their nuclei disintegrate to form an isotope of a different element. They may also emit very high-energy gamma (γ) radiation. When a nucleus emits an α-particle (= He nucleus of 2 neutrons and 2 protons) the atomic number decreases by 2 and the mass number decreases by 4. When a nucleus emits a β-particle (= an electron e^-) the atomic number increases by 1 but the mass number stays the same. Disintegrations may be represented by nuclear equations – for instance:

$$^{235}_{92}U \rightarrow \ ^{231}_{90}Th \ + \ \alpha^{2+} \quad \text{uranium decaying to thorium}$$

$$^{225}_{88}Ra \rightarrow \ ^{225}_{89}Ac \ + \ e^- \quad \text{radium decaying to actinium}$$

The radioactive decay of an isotope is a **first-order reaction**. The time taken for the radioactivity of a radioisotope to fall to half of its initial value is called the **half-life**. This period or half-life is a constant which is characteristic of the radioisotope. It is independent of the mass of the isotope and unaffected by **catalysts** or by changes in temperature. The shorter the half-life, the more unstable the radioisotope.

◀ Radioactivity ▶

ISOTROPIC CRYSTAL

An isotropic substance is a crystalline solid whose refractive index does not vary with the direction of the light through its **crystal**. Caesium chloride and sodium chloride are typical isotropic ionic crystalline compounds belonging to the cubic crystal system. Their structures are composed of spherical **ions** and do not rotate the plane of **polarised light**.

JOULE

The joule, symbol J, is the systematic international unit of energy equal to the work of moving a force of 1 newton a distance of 1 metre. The unit is named after the scientist James Prescott Joule (1818–89), a wealthy Manchester brewer's son whose experimental research established that heat and work are interconvertible forms of energy. For many purposes in chemistry the joule is too small a unit. The kilojoule (one thousand joules), symbol kJ, is more convenient. The k must be lower-case (NEVER upper-case K) and the J, must be upper-case (NEVER lower case j).

◀ Enthalpy change, Hess's law ▶

JUNCTION POTENTIAL

A liquid junction potential is a potential which can exist between two different **aqueous electrolytes** in contact with each other. If two **half-cells** are put in electrolytic contact, a potential difference can develop at the liquid junction, and this will affect the e.m.f. of the electrochemical cell. If the two half-cells are connected by a **salt bridge**, the liquid junction potential will be minimised.

◀ Electrochemical cell ▶

KELVIN

The kelvin, symbol K, is the systematic international unit of thermodynamic (absolute) temperature. On the kelvin scale, absolute zero corresponds to −273.15°C. On the Celsius scale, 0°C corresponds to 273.15K. For A-level chemistry calculations, a temperature can be converted from degrees centigrade to kelvin (NOT 'degrees kelvin') simply by adding 273. So 25°C becomes 298K.

The unit is named after Lord Kelvin, who was born in 1824 in Belfast, became a professor at Glasgow University at the age of 22, and remained there all his life.

KETONES

Ketones are a class of compounds containing the **carbonyl functional group** $> C=O$, which is never attached directly to a hydrogen atom.

 NOMENCLATURE

Ketones are isomeric with **aldehydes**. The names of ketones end with the suffix **-one** and a number to show the position of the carbonyl group in the chain. For example:

$$CH_3CH_2CH_2COCH_3$$
pentan-2-one

$$CH_3CH_2COCH_2CH_3$$
pentan-3-one

 PHYSICAL PROPERTIES

The lower ketones are typical **polar** covalent liquids with distinctive **infra-red absorption spectra** caused by the $>C=O$ group. Permanent **dipole−dipole attractions** but NOT hydrogen-bonding contribute to the **intermolecular forces** which make their melting and boiling points higher than those of **aliphatic hydrocarbons** and **ethers** but lower than those of aliphatic **alcohols** of similar molar mass. Propanone (acetone) forms **hydrogen bonds** with water and dissolves, so it is often used to rinse apparatus prior to drying. It is also a good solvent for many organic reactions because it will dissolve both ionic and covalent substances.

▶ *CHEMICAL PROPERTIES*

Ketones should be kept away from naked flames because they are dangerously flammable, forming explosive mixtures with air or oxygen. They are NOT oxidised by hot acidified aqueous sodium dichromate(VI) but some cyclic ketones degrade to a mixture of **carboxylic acids**, often with fewer carbon atoms than the original ketone, when refluxed with vigorous **oxidising agents** such as acidified potassium manganate(VII). Unlike most aldehydes, ketones do NOT reduce **Fehling's solution**

A hydrogen atom on the α carbon atom (the carbon atom next to the C=O group) of a ketone is readily replaced by a halogen atom. The acid-catalysed reaction of propanone with iodine is **first-order** with respect to propanone and to hydrogen ion but **zero-order** with respect to iodine:

$$CH_3-\underset{O}{\underset{\|}{C}}-CH_3 + I_2(aq) \xrightarrow{\quad H^+(aq) \quad} CH_3-\underset{O}{\underset{\|}{C}}-CH_2I + H^+(aq) + I^-(aq)$$

Propanone and other methyl ketones give a positive result (a yellow precipitate of CHI_3) with the iodoform test because they contain the $CH_3-C=O$ group. The alkyl groups in ketone molecules may underrgo **free-radical substitution reactions** with chlorine in ultraviolet light and the oxygen in the carbonyl group may be replaced by chlorine by reaction with phosphorus pentachloride under anhydrous conditions; for instance:

$$(CH_3)_2CO \xrightarrow[\text{conditions}]{\text{PCl}_5 \text{ in anhydrous}} (CH_3)_2CCl_2 \quad \text{2,2-dichloropropane}$$

Ketones are reduced to **secondary alcohols** by, for example, sodium tetrahydridoborate ($NaBH_4$) or by lithium tetrahydridoaluminate ($LiAlH_4$) in dry ethoxyethane. The **reduction** may be seen as i) **nucleophilic addition** of hydride anions, $:H^-$, supplied by BH_4^- and AlH_4^- ions, to the electrophilic carbon atom in the $>C=O$ group followed by ii) electrophilic attack of hydrogen cations, H^+, supplied by the water molecules, upon the nucleophilic oxygen atom in the carbonyl group.

Ketones undergo nucleophilic addition reactions with HCN as NaCN(aq) followed by HCl(aq) and with $NaHSO_3$(aq) or Na_2SO_3(aq). They also undergo addition–elimination reactions with derivatives of ammonia such as hydroxylamine. These are the typical reactions of carbonyl compounds. In contrast to aldehydes, ketones do not polymerise to form resins.

Propanone is an industrially important ketone manufactured as a co-product of the Cumene process for the production of **phenol**.

KINETICS

Kinetics is the branch of **physical chemistry** concerned with the study of reaction rates and the influence on them of such factors as **concentration**, pressure, temperature and **catalysts**.

KINETIC THEORY

The kinetic theory explains the physical properties of gases in terms of the continuous random motion of molecules, and it derives an expression for the pressure, p, and volume, V, of an ideal gas in terms of the number, N, mass, m, and mean square speed, $\bar{c}^2$ of its molecules: $pV = \frac{1}{3} Nm\bar{c}^2$.

The kinetic molecular theory makes the following important assumptions. 1) The collisions of the molecules with the walls of the container and with one another are perfectly elastic so the total kinetic energy of all the molecules of gas is constant. 2) The average kinetic energy of the molecules is directly proportional to the temperature of the gas measured on the Kelvin scale. 3) The forces of attraction between the molecules, and between the molecules and the walls of the container, are negligible. 4) The actual volume of a gas molecule, and therefore of all the gas molecules, is negligible compared to the volume of their container.

The kinetic molecular theory can account for Boyle's law, Charles' law, the ideal gas law and Graham's law of diffusion.

◀ Gas laws ▶

LAEVOROTATORY

A laevorotatory substance is an optically active substance which rotates the plane of **polarised light** to the left in an anticlockwise direction, viewed from a position facing the plane-polarised light source.
◄ **Optical activity** ►

LANTHANOIDS

The lanthanoids (*lanthanides* or *rare earths*) are a horizontal row of elements from cerium to lutetium in the **f-block** of the **periodic table**.

LATTICE

Lattice is a term which describes the regular arrangement of **atoms**, **ions** or molecules in a **crystal**.
◄ **Crystal lattice** ►

LATTICE ENERGY

Lattice energy may be defined as the heat given out when one **mole** of an **ionic** **crystalline** solid such as sodium chloride forms from its constituent **ions** in the gaseous state: $Na^+(g) + Cl^-(g) \rightarrow NaCl(s)$.

If the lattice energy is considered as heat released at constant pressure, it should be called the molar lattice **enthalpy change**. If the lattice energy is considered as heat released at constant volume, it should be called the molar lattice **internal energy change**.

► EXPERIMENTAL LATTICE ENERGIES

Lattice energies may be determined from a **Born–Haber** cycle using **Hess's law**. Values range from around $600 kJ\ mol^{-1}$ (beyond the strength of single covalent bonds) to above $4000\ kJ\ mol^{-1}$ (well beyond the strongest covalent bonds). Two important patterns in the values show the influence of the charge density of the ions upon lattice energy: i) Lattice energies for doubly charged ions are usually much more **exothermic** than those for singly charged ions; ii) Lattice energies become less exothermic as the size of the ions increases. Fig. L.1 shows patterns in lattice energies.

Lattice	Energy /kJ mol^{-1}	Lattice	Energy /kJ mol^{-1}	Lattice	Energy /kJ mol^{-1}
Na^+Cl^-	−787	LiF	−1031	NaF	−918
$(Na^+)_2O^{2-}$	−2478	LiCl	−848	KF	−817
$Mg^{2+}(Cl^-)_2$	−2526	LiBr	−803	RbF	−783
$Mg^{2+}O^{2-}$	−3791	LiI	−759	CsF	−747

Fig. L.1

▶ THEORETICAL LATTICE ENERGIES

Completely theoretical lattice energies have been calculated on the assumption that each metal **atom** loses completely one (or two) **electrons** to form a perfectly spherical **cation** and that each non-metal atom gains completely one (or two) electrons to form a perfectly spherical **anion**. The experimental and theoretical lattice energies for the **alkali metal** halides are *similar* and indicate that these compounds form **ionic crystals** with little or no covalent character. Those for the silver halides are *dissimilar* and indicate that these compounds have considerable covalent character.

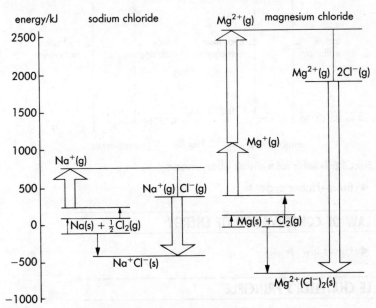

Fig. L.2 Lattice energy compensates for ionisation energy

▶ *LATTICE ENERGY AND COMPOUND FORMATION*

The major contribution to the **enthalpy of formation** of a compound may be seen as the difference between the energy consumed by the ionisation of the metal to gaseous cations and the energy released by the formation of the lattice (Fig. L.2). If the **ionisation energy** taken in would be much greater than the lattice energy given out, the formation of the compound would be **endothermic** and could be too unstable to exist. For example, the enthalpy of formation of $MgCl_3$ would be about $+4000$ kJ mol^{-1} because the ionisation of Mg^{2+} to Mg^{3+} involves a new inner electron shell and requires about 7700 kJ mol^{-1} of energy.

▶ *LATTICE ENERGY AND ENTHALPIES OF SOLUTION*

The enthalpy of solution of an ionic compound may be regarded as the difference between the lattice energy of the crystal and the **hydration** energy of the ions. The solution process may be exothermic or endothermic. The heat change is usually quite small because it is the difference between two large energy values (see Fig. L.3).

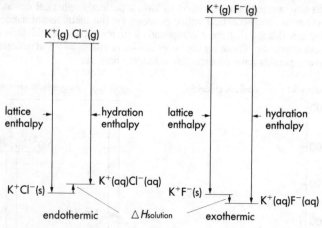

Fig. L.3 Endothermic and exothermic solution processes

◀ Born–Haber cycle ▶

LAW OF CONSERVATION OF ENERGY

◀ Hess's law ▶

LE CHATELIER'S PRINCIPLE

In 1888 Henri Le Chatelier, a French physical chemist, put forward a principle to describe the effects of stress upon a system in equilibrium. There are probably as many different translations as there are textbooks; here is a version:

If you alter the conditions of a **reversible reaction** and disturb the equilibrium, the composition of the mixture may change to restore the equilibrium and to minimise the effect of altering the conditions.

Le Chatelier's principle deals *qualitatively* with the way the composition of an equilibrium system may change with temperature, **concentration** and pressure. The **law of chemical equilibrium** and the expression for the **equilibrium constant** (K_c or K_p) deal *quantitatively* with these effects.

For example, according to Le Chatelier's principle, because the formation of ammonia is accompanied by a decrease in volume, the proportion of ammonia in the equilibrium mixture of nitrogen, hydrogen and ammonia at a constant temperature should increase with an increase in pressure.

$$N_2(g) + 3H_2(g) \rightleftharpoons 2NH_3(g); \quad K_P = 41 \text{ atm}^{-2} \text{ at } 127°C$$

When applying Le Chatelier's principle to a reversible reaction at equilibrium, a common mistake is to think that the value of the equilibrium constant changes with a change in concentration or pressure. It does not. The value of K_c or K_p changes with temperature and with the nature of the reaction. K_c and K_p for a given reaction at a constant temperature does NOT change with concentration or with pressure.

◀ Equilibrium, Equilibrium constant, Haber process ▶

LEWIS ACID

A Lewis acid is a **molecule** or **ion** that can accept a pair of **electrons** to form a **covalent bond**. This broad definition of an acid was proposed in 1923 by Gilbert Newton Lewis, an American professor of chemistry, in the same year that Johannes Bronsted and Thomas Lowry put forward their definition of an acid as a proton donor. Bronsted–Lowry acids are included in the Lewis definition because every Bronsted–Lowry acid will donate a proton that accepts a **lone pair of electrons** to form a **dative-covalent bond** with a base – for example:

$$CH_3CO_2H \rightarrow CH_3CO_2^- + H^+ \quad \text{then} \quad H^+ + :OH^- \rightarrow H:OH$$

However, the Lewis definition will also include an acid such as boron trichloride, which reacts with a base such as ammonia, even though the acid does not donate a proton: $BCl_3 + :NH_3 \rightarrow Cl_3B:NH_3$

◀ Acids, Bronsted–Lowry theory ▶

LEWIS BASE

A Lewis base is a **molecule** or **ion** which can donate a pair of **electrons** to form a **covalent bond** with a **Lewis acid**. $:NH_3$, $H_2O:$, $:CN:^-$ and $:OH^-$ are Lewis bases. However, they are called **ligands** when they form **dative bonds** in metal **complexes**, and **nucleophiles** when they attack electrophilic centres in organic **reaction mechanisms**.

◀ Hydrolysis, Nucleophilic substitution, Reaction mechanism ▶

LIGAND

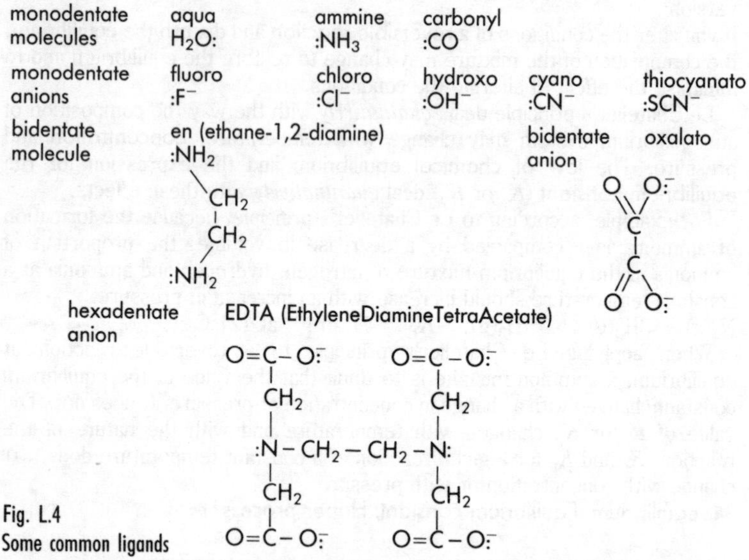

monodentate molecules	aqua $H_2O:$	ammine $:NH_3$	carbonyl $:CO$		
monodentate anions	fluoro $:F^-$	chloro $:Cl^-$	hydroxo $:OH^-$	cyano $:CN^-$	thiocyanato $:SCN^-$

bidentate molecule   en (ethane-1,2-diamine)

$:NH_2$
$\quad \backslash CH_2$
$\quad\quad CH_2$
$:NH_2$

bidentate anion   oxalato

hexadentate anion   EDTA (EthyleneDiamineTetraAcetate)

Fig. L.4
Some common ligands

A ligand is an atom, or group of atoms, dative-covalently bonded to a central **transition metal** atom in a **complex**. Before they become bonded into a complex, ligands are molecules or **anions** with one or more atoms having a **lone pair** of electrons that can be donated to form a dative bond with the central atom or ion in the complex. A *monodentate* (one-tooth) ligand forms one dative bond with the central ion. A **bidendate** (two-tooth) ligand consists of several atoms, two of which are each able to form one dative bond with the central ion. A **hexadentate** ligand consists of many atoms, six of which each form one dative bond with the central ion.

◄ Complex ►

LINE EMISSION SPECTRUM

A line emission spectrum is a spectrum, consisting of a series of coloured lines. This is produced by light from an incandescent gas or from metal chlorides heated in a bunsen flame.

◄ Atomic emission spectrum, Lyman series ►

LONE PAIR

A lone pair is the term for two unshared **valence electrons** in an **atom**. In an ammonia molecule, the nitrogen atom has three bonded electron pairs, each shared with a separate hydrogen atom, and one non-bonded electron pair:

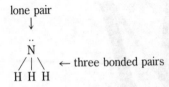

In a water molecule, the oxygen atom has two bonded electron pairs and two lone pairs

The **bond angles** and shapes of these and other simple molecules may be explained by the decreasing strength of the repulsions between two lone pairs, a lone pair and a bonded pair, and two bonded pairs. The lone pairs allow the above molecules to form **dative bonds** with other atoms and to act as Lewis bases, ligands and nucleophiles.

◀ Valence shell electron pair repulsion theory (VSEPR) ▶

LOWERING OF THE VAPOUR PRESSURE

The lowering of the vapour pressure ($\triangle p$) of a liquid is the difference between the vapour pressure ($p°$) when the liquid is pure and the vapour pressure (p) when the liquid is the solvent in a solution: $\triangle p = p° - p$

The *relative* lowering of the vapour pressure of a solvent, $\triangle p/p°$ is equal to the **mole fraction** of the solute in **ideal dilute solutions** of non-electrolytes – so $\triangle p/p° = n_A/(n_A + n_s)$
where n_A and n_s are the amounts of solute and solvent respectively.

Vapour pressure lowering is a **colligative property** that can be used to determine **molecular weights**.

◀ Colligative property, Raoult's law ▶

LYMAN SERIES

The Lyman series is the series of lines in the ultraviolet region of the **atomic emission spectrum** of hydrogen whose wavelengths (λ) are given by the formula $1/\lambda = R\{(1/n^2) - (1/m^2)\}$ where R is the Rydberg constant (not the gas constant), $n = 1$ and $m = 2, 3, 4$, etc.

The series of lines converges to a limit at a wavelength of approximately 91.2 nm. The energy of the ultraviolet radiation with this wavelength corresponds to the energy needed to ionise a hydrogen **atom** from its ground state. This **ionisation energy** of hydrogen is the energy to move an **electron** from the principal quantum energy level $n = 1$ to an infinite distance from tthe nucleus.

◀ Atomic emission spectrum ▶

MACROMOLECULE

A macromolecule is a very large molecule. The term is usually applied to organic molecules of very high **molar mass**. Some organic substances, such as certain **proteins**, consist of giant molecules with a unique composition and formula for each protein; for example, insulin has been given the molecular formula $C_{254}H_{377}N_{65}S_6$. Natural and synthetic organic **polymers** are macromolecular substances whose giant molecules vary in size according to the number of repeating monomer units in the polymer. The structures of some elements, such as carbon, silicon and phosphorus, are described as **giant molecular structures** because all the atoms are covalently bonded to each other in a three-dimensional network to give one giant molecule.

◀ Giant structure, Polymers ▶

MARKOVNIKOV'S RULE

When a molecule HZ undergoes an **electrophilic addition** across a $C=C$ **double bond** in an unsymmetrical **alkene**, the hydrogen adds to the carbon atom with the greater number of H atoms already attached.

The addition reaction between hydrogen bromide and propene normally yields 2-bromopropane and NOT 1-bromopropane:

$$HBr(g) + CH_3CH=CH_2(g) \rightarrow CH_3CH-CH_3(l) \quad \text{2-bromopropane}$$
$$|$$
$$Br$$

The rule is NOT obeyed by **free radical** addition reactions occurring in the presence of peroxides or ultraviolet light.

◀ Electrophilic addition reaction ▶

MASS NUMBER

Mass number (A) is the number of **protons** (Z) and **neutrons** in the **nucleus** of a nuclide. The mass number of an **isotope** of an element is shown as a superscript to the symbol: ^{35}Cl and ^{37}Cl represent, for instance, the chlorine isotopes of mass number 35 and 37 respectively.

MASS SPECTROMETER

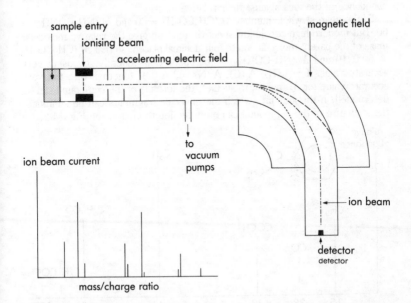

Fig. M.1

A mass spectrometer (Fig. M.1) is essentially an analytical instrument which involves the production, separation and detection of **ions** at extremely low pressures. The earliest instrument recognisable as a mass spectrometer was produced by A.J. Dempster in Chicago in 1918. Mass spectrometers have been used to accurately measure the **relative atomic masses** of elements by determining the abundances of their **isotopes**. More recently, valuable information has been obtained by determining the isotopic abundances in a mineral sample: the abundancies of Pb-206, Pb-207 and Pb-208 (end-products of naturally occurring radioactive decay series), for example, may indicate the presence of economic quantities of uranium and thorium for mining.

The most widespread use of mass spectrometers is the determination of **relative molecular masses** (M_r), **molecular formulae** and structures of compounds. The substance is vaporised in the instrument at extremely low pressure. The gaseous molecules are ionised and usually fragmented into smaller gaseous ions. A fine beam of these ions (having different masses) is accelerated by an electric field and deflected by a magnetic field on to a (positive ion) current detector. The magnitude of this ion-beam current registered by the detector depends upon the number of ions in the beam focused on the detector. The mass of the ions being detected at any time depends upon the settings of the electric and magnetic fields. Increasing the magnetic field strength enables ions of increasingly higher mass to be focused and detected.

A **mass spectrum** may be a chart recording of ion-beam current against magnetic field strength (see Fig. M.1). It may be regarded as a plot of the abundance of the ions against their relative masses.

To the nearest whole number $M_r(CH_3CO_2H) = 60$ and $M_r(NH_2CONH_2) = 60$. But modern high-**resolution** spectroscopy can give the relative molecular mass of the parent molecule ion to four decimal places, so that $M_r(CH_3CO_2H) = 60.0210$ and $M_r(NH_2CONH_2) = 60.0323$. Consequently, by using very accurate values for $A_r(H)$, $A_r(C)$, $A_r(N)$ and $A_r(O)$, together with the very accurate value for M_r(parent molecule), the formula of the compound can be determined. If the ions into which the molecule fragments can be identified, the structure of the compound can often be determined – see Fig. M.2.

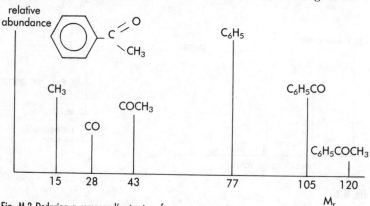

Fig. M.2 Deducing a compound's structure from a mass spectrum

◀ Molecular spectroscopy, Structural formula ▶

MASS SPECTRUM

A mass spectrum (Figs. M.1 and M.2) is usually the ion-beam current against mass/charge ratio display from a **mass spectrometer** which shows the fragmentation pattern of the sample introduced into the mass spectrometer.

MAXWELL–BOLTZMANN DISTRIBUTION

The Maxwell–Boltzmann distribution is the spread of speeds and energies in a collection of **molecules**. James Clerk Maxwell (1831–79) derived, and Ludwig Boltzmann (1844–1906) rigorously proved, a mathematical expression for the distribution of molecular velocities as an exponential function of their kinetic energies. The distribution of the energies is usually displayed in the form of a graph like the one in Fig. A.11.

The function $f(E, n)$ makes the total area under the curve proportional to the total number of molecules and the area under any part of the curve proportional to the number of molecules with energies in that range. Consequently, as the temperature increases, i) the peak of the curve moves to the right so the mean value of the function $f(E, n)$ (and therefore the mean energy of the molecules) increases; ii) the curve flattens so the total area under it (and therefore the total number of molecules) remains constant; and iii) the area under the curve to the right of E_a, the **activation energy**, roughly doubles for every 10-degree rise in temperature.

◀ Activation energy, Kinetic theory, Rate of reaction ▶

MECHANISM

◀ Reaction mechanism ▶

MELTING POINT

The melting point of a substance is the temperature at which the solid and liquid phases can coexist in **equilibrium** at a standard pressure of 1 atm. The melting point is the same as the **freezing point** of a substance, but it is the name usually used when the substance is in the solid phase and is being heated up to form the liquid phase.

◀ Freezing point ▶

MENDELEYEV, DMITRI IVANOVIC

◀ Periodic law, Periodic table ▶

METALLIC BONDING

Metallic **bonding** is the bonding in a metal structure which may be described as a lattice of mutually repelling positive ions held together by their attraction for a 'sea' of mobile delocalised electrons.

Metallic bonding is non-directional and strong, the strength arising from the very powerful attraction between the positive ions and the electrons. The mobility of the electrons and the non-directional nature of the attractive forces account for properties such as lustre, high electrical and thermal conductivity, hardness, toughness, ductility and malleability, high melting points, high boiling points and high molar enthalpies of melting and boiling that are characteristic of metals. The usually stronger metallic bonding in **d-block metals** by comparison with **s-block metals** is related to the ability of the d-block metal atoms also to contribute bonding electrons from their d-orbitals.

◀ Bonding, Delocalisation ▶

MINERAL ACID

The mineral acids are the common inorganic acids: hydrochloric acid HCl(aq) nitric acid HNO_3(aq) and sulphuric acid H_2SO_4(aq). These acids may be called mineral acids because they were probably first prepared by chemical reactions involving minerals such as rock salt, saltpetre and pyrite.

◀ Acids ▶

MIXED MELTING POINT

A mixed melting point determination is an analytical technique to confirm the identity of a solid by mixing it with a pure sample of a known substance and measuring the melting point of the mixture. If two pure samples of the same substance are mixed together, the melting point of the mixture will be sharp and the same as that of the separate samples. If two pure samples of different substances, are mixed together, the melting point of the mixture will not be sharp and not be the same as that of the separate substances.

◀ Qualitative analysis ▶

MOLALITY

The molality is the **concentration** of a solution expressed in **moles** of solute per kilogram of solvent. Unlike the **molarity** of a solution, the molality does not change with temperature.

◀ Concentration ▶

MOLARITY

Molarity is a widely used term for the **concentration** of a solution expressed in moles of solute per cubic decimetre of solution. Aqueous sulphuric acid containing 0.1 mol H_2SO_4(aq) in 1 dm^3 of solution is often said to have a molarity of 0.1 and described as a 0.1 molar solution or 0.1 M H_2SO_4(aq). The term molarity is best avoided, and should not be confused with the term molality.

◀ Concentration ▶

MOLAR MASS

Molar mass, symbol M, has the units g mol^{-1} and is the mass of one **mole** of substance specified by its chemical formula. It should not be confused with **relative molecular mass**, symbol M_r, or with the obsolete term **molecular weight**.

◀ Mole ▶

MOLAR VOLUME

Molar volume is the volume occupied by one **mole** of substance under specified conditions. The molar volume of an ideal gas is approximately 22.4 dm^3 at 273 K and 1 atm. The molar volume of most gases is about 24 dm^3 at 298 K and 1 atm.

MOLE

The mole is the amount of substance which contains as many entities (**atoms, molecules, ions,** etc.) specified by the **formula** as there are carbon atoms in 0.012 kg of carbon nuclide ^{12}C.

The mole has replaced the now obsolete units of gram-atom, gram-molecule and gram-formula as the unit (symbol mol) for an amount of substance (symbol n). An expression such as n(Mg) = 2.0 mol means 'the amount of magnesium atoms in two moles'. The number of entities (atoms, molecules, ions, electrons, etc.) in one mole is 6.022×10^{23} (sometimes called the Avogadro number). An entity should always be unambiguously specified by its formula. A given amount of substance, say n(Mg), is multiplied by the **Avogadro constant** (L) to calculate the number of entities (N magnesium atoms), thus: N = $L \times n$(Mg).

The mass of one mole of substance (the molar mass) can be calculated using the formula for the entity and the A_r values of the constituent elements. For example, since the formula of water is H_2O and A_r(H) = 1.0 and A_r(O) = 16.06, $M_r(H_2O)$ = 18.0, so the mass of one mole of H_2O would be 18 g. The molar masses (in g mol^{-1}) and formulae of many substances are listed in data books.

◀ Amount of substance, Avogadro constant, Concentration, Molar mass ▶

MOLECULAR FORMULA

A molecular formula shows the number of **atoms** of each element in a **molecule** or molecular unit and is an integral multiple of the **empirical formula** of a substance (see Fig. F.6).

◀ Formula ▶

MOLECULARITY

The molecularity of a step in a reaction is the number of species (**atoms, ions** or **molecules**) forming the **activated complex** in the **transition state** of that particular reaction step.

A reaction step is said to be **unimolecular, bimolecular** or, sometimes **trimolecular** if the number of reactant molecules is 1, 2 or 3 respectively. When the **rate of a reaction** is governed by a slow reaction step, the overall reaction is sometimes described in terms of the molecularity of that step. For example, alkaline hydrolysis reactions of halogenoalkanes may be described as S_N1 or S_N2 reactions – that is, nucleophilic substitutions which are unimolecular and bimolecular respectively.

The molecularity of a reaction step should not be confused with the order of a reaction overall.

◀ Order of reaction, Reaction mechanism ▶

MOLECULAR SPECTROSCOPY

Molecular spectroscopy is a term which refers to a set of analytical techniques involving the measurement of electromagnetic radiation absorbed or emitted by energy changes in the molecules of a substance.

◀ Infra-red absorption spectroscopy, Nuclear magnetic resonance, Ultraviolet absorption spectroscopy ▶

MOLECULAR WEIGHT

Molecular weight is the obsolete term for relative molecular mass.

MOLECULE

A molecule is a group of covalently bonded atoms forming the smallest identifiable unit responsible for the properties of a substance. The noble gases and metals in the gaseous state form *monatomic* molecules. The halogens and gaseous elements such as oxygen and nitrogen form diatomic molecules. The solid non-metallic elements form *polyatomic* and *giant* molecules. The composition of a molecule is shown by the molecular formula.

MOLE FRACTION

The mole fraction (x_A) of a component (A) in a mixture is its amount (n_A) divided by the total amount ($n_A + n_B + n_C$, etc.) of components in the mixture: $x_A = n_A/(n_A + n_B + n_C$, etc).

A mole fraction may have any value from 0 to 1 inclusive. The sum of the mole fractions of all the components in a mixture is $x_A + x_B + x_C$, etc. $= 1$.

The mole fraction of component in a solution of volatile liquids connects the partial vapour pressure of the component to its pure vapour pressure in Raoult's law: $p_A = x_A \times p_A^0$.

MONOBASIC ACID

A monobasic (or *monoprotic*) acid means an acid such as nitric acid (HNO_3) with one mole of donatable protons per mole of the acid.

One mole of a monobasic acid will react with one mole of a (monacidic) base such as sodium hydroxide to form the normal salt:

$$HNO_3(aq) + NaOH(aq) \rightarrow H_2O(l) + NaNO_3(aq) \text{ sodium nitrate}$$

Ethanoic acid (CH_3CO_2H) is monobasic, NOT tetrabasic, because the three hydrogen atoms attached to the carbon atom in the methyl group cannot be donated as protons. The proton is donated by the carboxyl group in the ethanoic acid molecule:

$$CH_3-C\overset{\displaystyle O}{\underset{\displaystyle OH(aq)}{}} \quad H_2O(l) \rightleftharpoons CH_3-CO_2^-(aq) + H_3O^+(aq)$$

Although hydrogen fluoride would be expected to be a strong monobasic acid like the other hydrogen halides, hydrofluoric acid is weak and forms exceptionally strongly **hydrogen-bonded dimers**, which act as dibasic acid molecules capable of forming an acid salt such as KHF_2. According to the Bronsted–Lowry definition of an acid, all acids are monobasic.
◀ Acid, Bronsted–Lowry theory ▶

MONOHYDRIC ALCOHOL

A monohydric alcohol is an alcohol with only one **hydroxyl group** (OH) in its molecular structure. The **straight-chain** monohydric primary alcohols form a **homologous series**, the first four members of which are methanol CH_3OH, ethanol CH_3CH_2OH, propan-1-ol $CH_3CH_2CH_2OH$ and butan-1-ol $CH_3CH_2CH_2CH_2OH$. Alcohols with more than one hydroxyl group per molecule are *dihydric*, *trihydric*, etc., or *polyhydric* alcohols.
◀ Alcohol ▶

MONOMER

A monomer is a compound, or one of its **molecules**, capable of combining with itself or another compound to form a **polymer**. **Alkenes** such as ethene $CH_2=CH_2$, propene $CH_3CH=CH_2$ and phenylethene $C_6H_5CH=CH_2$ are the monomers of the poly(alkene) addition polymers. **Amino acids**, monosaccharides, dicarboxylic acids and dihydric alcohols are the monomers of the **condensation polymers** called **proteins, polysaccharides** and **polyesters** respectively.
◀ Polymer ▶

MONOPROTIC ACID

◀ Monobasic acid ▶

MOSELEY'S EXPERIMENTS

In Henry G. Moseley's 1913 experiments, elements bombarded by high-energy electrons emitted X-rays whose frequencies (v) were found to be related to the **atomic number** (Z) of the elements: $\sqrt{v} = a(Z - b)$ where a and b are constants.

Prior to Moseley's work, the atomic number of an element was an ordinal number referring to the position of the element in a **periodic table** and in a list arranged in order of increasing **atomic weight**. When Moseley accurately measured the frequencies of the two intense components (K and L lines) of the X-rays emitted by various metallic elements and plotted the square root of these frequencies against atomic number, he obtained straight-line graphs – see Fig. M.3.

Moseley's work was taken as showing that the atomic number of an element is a cardinal number referring to the number of protons in the nucleus of an atom of the elements, and his graphs provided a way of measuring it.

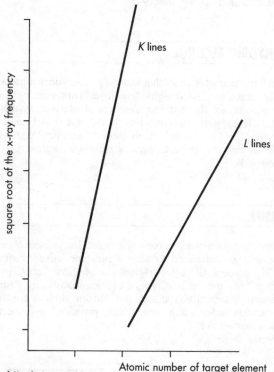

Fig. M.3 Results of Moseley's experiments

◀ Atomic number ▶

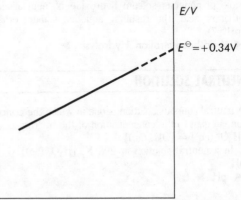

NERNST EQUATION

Walther Nernst (1864–1941) was a German physical chemist noted for his work in electrochemistry and his development of the third law of thermodynamics. He won the Nobel Prize in 1920. A Nernst equation is a mathematical expression for the e.m.f. (E_{cell}) of an electrochemical cell in terms of the concentrations of the reactants (A and B) and products (C and D) of the cell redox reaction $aA + bB \rightarrow cC + dD$:

$$E_{cell} = E_{cell}^{\theta} + \frac{RT}{zF} \ln \frac{[A]^a[B]^b}{[C]^c[D]^d}$$

where R is the **gas constant**, F is the **Faraday constant**, z is the number of moles of electrons transferred from reductant to oxidant per mole of reaction and E_{cell}^{θ} is the **standard e.m.f.** of the cell at a temperature of T kelvin.

When the cell e.m.f. becomes zero, the cell **redox reaction** comes to **equilibrium** and the **concentrations** of the reactants and products have their equilibrium values. If $E_{cell} = 0$, it follows from the Nernst equation that $E_{cell}^{\theta} = (RT/zF)\ln K_c$, where K_c is the **equilibrium constant** for the cell redox reaction.

E/V

$E^{\ominus} = +0.34V$

$\ln[Cu^{2+}(aq)]$

Fig. N.1 Graph of electrode potential against the logarithm of the copper(II) ion concentration

For the electrochemical cell $Pt[H_2(g)]\ 2H^+(aq) :: Cu^{2+}(aq)\ |\ Cu(s)$ in which the left-hand half-cell is the standard hydrogen electrode, the Nernst equation will describe the dependence of the electrode potential of the $Cu^{2+}(aq)\ |\ Cu(s)$ system upon the concentration of the $Cu^{2+}(aq)$ ion:

$$E = E^\theta + \frac{RT}{2F} \ln[Cu^{2+}(aq)]$$

This equation shows that a graph of the electrode potential (E) plotted against $\log[Cu^{2+}(aq)]$ should be a straight line, giving a value for the standard electrode potential of the $Cu^{2+}(aq)\ |\ Cu(s)$ system when extrapolated to $\ln[Cu^{2+}(aq)] = 0$ that is, when $[Cu^{2+}(aq)] = 1$ mol dm^{-3} (see Fig. N.1).

◀ Electrochemical cell, Electrode potential, Standard electrode potential ▶

NEUTRALISATION

Neutralisation is the reaction of an acid and base usually, but not always, to form water and a solution with a pH of 7.

Aqueous strong acids react with aqueous strong bases to form neutral aqueous salts and water. When one mole of water is formed by the reaction of strong acids and bases, the enthalpy change of about -57 kJ mol^{-1} is the molar enthalpy change of neutralisation for the reaction:

$$H_3O^+(aq) + OH^-(aq) \rightarrow 2H_2O(l)$$

Less heat is usually evolved in the neutralisation of weak acids and bases because weak acids and bases are only partially ionised in aqueous solution. If a mole of aqueous ammonia is 'neutralised' by a mole of aqueous hydrochloric acid, the resulting aqueous ammonium chloride is slightly acidic (pH<7). If a mole of aqueous sodium hydroxide is 'neutralised' by a mole of aqueous ethanoic acid, the resulting aqueous sodium ethanoate is slightly alkaline (pH>7).

◀ Acid-base titration, Hydrolysis ▶

NEUTRAL SOLUTION

A neutral aqueous solution is one in which the concentration of the hydrogen ion is equal to the concentration of the hydroxide ion: that is, $[H_3O^+(aq)] = [OH^-(aq)]$.

In a neutral solution at 298 K, $[H_3O^+(aq)] = 1 \times 10^{-7}$ mol dm^{-3} and pH = 7.

◀ pH ▶

NEUTRON

A neutron (n) is an uncharged sub-atomic particle with a mass of

1.675×10^{-27} kg. The mass of a neutron is just slightly greater than the mass of a **proton**. Neutrons together with protons are called nucleons. They constitute the **nucleus** and most of the mass of an **atom**.

Different numbers of neutrons in the nucleus are responsible for the different mass numbers of nuclides and the existence of isotopes. Free neutrons are emitted by fission of an unstable nucleus of, say, uranium (^{235}U) or plutonium (^{239}Pu). These neutrons have a high penetrating power and are radioactive, decaying into a proton, an electron and an antineutrino. Free neutrons can cause further nuclear fission and, if uncontrolled, can lead to an explosive nuclear chain reaction. Fortunately, elements such as cadmium can readily absorb free neutrons and act as control rods in a nuclear reactor.

◀ Mass number ▶

NEUTRON NUMBER

The neutron number is the number of neutrons in a **nucleus**.

NITRATE

Nitrates are the salts of nitric acid $HNO_3(aq)$. As a general rule, all nitrates are soluble in water. Large deposits of sodium nitrate occur, together with sodium iodate, as *caliche* in the arid desert region on the west coast of South America. Metal nitrates can be made by reacting the metal, its **oxide**, **hydroxide** or **carbonate** with nitric acid. Ammonium nitrate, an important fertiliser and explosive, is manufactured by the **oxidation** of ammonia to nitric acid, followed by the reaction of ammonia with nitric acid:

$NH_3 + HNO_3 \rightarrow NH_4NO_3$.

◀ Acid, Mineral acid ▶

NITRATION

Nitration is the **electrophilic substitution** of nitro groups ($-NO_2$) into organic compounds by reaction with nitric acid, often in the presence of concentrated sulphuric acid.

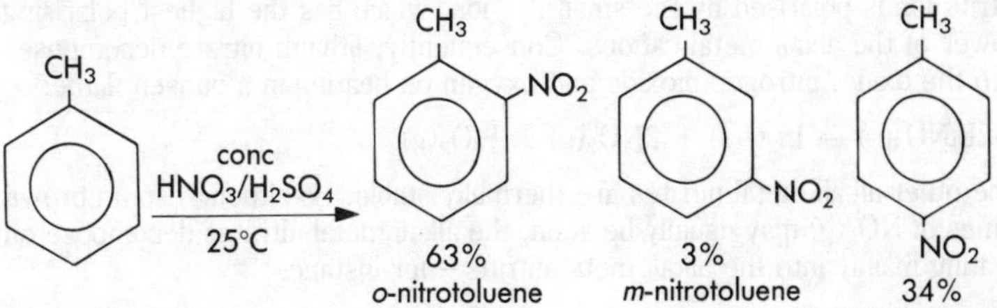

63%
o-nitrotoluene
or
2-nitromethylbenzene

3%
m-nitrotoluene
or
3-nitromethylbenzene

34%
p-nitrotoluene
or
4-nitromethylbenzene

Fig. N.2 Isomeric substition products of methylbenzene

A mixture of concentrated nitric acid and concentrated sulphuric acid provides the electrophilic nitronium ion (nitryl cation) which can attack the nucleophilic benzene ring: $HNO_3 + 2H_2SO_4 \rightarrow NO_2^+ + H_3O^+ + 2HSO_4^-$.

The products of a nitration reaction depend not only upon the organic compound being nitrated but also upon the conditions, such as temperature and concentration of the reagents. Nitration of methylbenzene takes place faster and at a lower temperature than for benzene itself; substitution occurs mainly at positions *ortho* and *para* to the methyl group (see Fig. N.2).

◄ Electrophilic substitution ►

NITRILES

Nitriles are organic compounds containing the **functional group** −CN. These toxic, dangerously flammable liquids are also known as cyanides (for instance, CH_3CN methyl cyanide or ethanenitrile). Nitriles can be synthesised by heating alcoholic potassium cyanide with a halogeno-compound. This **nucleophilic substitution** reaction is one way of precisely lengthening a carbon chain:

$$CH_3CH_2Br + K^+CN^- \xrightarrow[\text{alcoholic conditions}]{\text{reflux under}} CH_3CH_2CN + K^+Br^-$$

bromoethane propanenitrile

Nitriles are important intermediates in organic syntheses. They can be hydrolysed to **carboxylic acids** ($R-CN \rightarrow R-CO_2H$) and reduced to primary **amines** ($R-CN \rightarrow R-CH_2NH_2$). Nitriles can also be made by **dehydration** of **amides** ($R-CONH_2 \rightarrow R-CN$). Propenenitrile (acrylonitrile) is an important monomer in the industrial production of the **polymer** 'Acrilan'.

NITRITES

Nitrites are the salts of nitrous acid $HNO_2(aq)$.

Lithium nitrite is less stable than the other **alkali metal** nitrites, because the nitrite ion is polarised by the small Li^+ ion, which has the highest polarising power of the alkali metal cations. Consequently, lithium nitrate decomposes into the oxide, nitrogen dioxide and oxygen on heating in a bunsen flame:

$$2LiNO_3(s) \rightarrow Li_2O(s) + 2NO_2(g) + \tfrac{1}{2}O_2(g)$$

The other alkali metal nitrites are thermally stable. So although some brown fumes of $NO_2(g)$ may usually be seen, the alkali metal nitrates decompose on heating mainly into the alkali metal nitrite − for instance:

$$NaNO_3(s) \xrightarrow{\text{heat}} NaNO_2(s) + \tfrac{1}{2}O_2(g)$$

Nitrites are readily distinguished from **nitrates** by the action of ethanoic acid. The acid has no effect upon nitrates, but with nitrites it gives a pale blue solution of unstable nitrous acid spontaneously decomposing or disproportionating into a colourless gas which turns brown in air:

$$NaNO_2(aq) + CH_3CO_2H(aq) \rightarrow CH_3CO_2Na(aq) + HNO_2(aq)$$

$$3HNO_2(aq) \rightarrow H_2O(l) + HNO_3(aq) + 2NO(g)$$

$$NO(g) + \frac{1}{2}O_2(g) \rightarrow NO_2(g)$$

Care is needed not to confuse the formula of the nitrite ion, NO_2^-, with the formula of the nitrate ion, NO_3^-.

NMR

◀ Nuclear magnetic resonance ▶

NOBLE GASES

The noble gases – helium, neon, argon, krypton, xenon and radon – are a family of unreactive monatomic gases in Group O of the periodic table. They have very similar physical properties which show typical trends. Boiling point and atomic radius increase, and first ionisation energy decreases with increasing atomic number down the group.

The noble gases are generally unreactive, helium, neon and argon not seeming to form any compounds at all. The lack of reactivity is related to the electronic configurations of their atoms, in which all the electron shells are filled. Xenon will combine with fluorine to form xenon(IV) fluoride, whose square planar shape can be predicted by applying the valence shell electron pair repulsion (VSEPR) rules.

$$Xe(g) + 2F_2(g) \xrightarrow{\text{400°C 6 atm}} XeF_4(s)$$

The uses of the noble gases mainly depend upon their unreactive nature. For example, helium (instead of flammable hydrogen) is used for airships, neon for filling fluorescent lighting tubes and argon for an inert atmosphere in the high-temperature reduction of titanium(IV) chloride to titanium.

◀ p-block elements ▶

NOMENCLATURE

Nomenclature (Latin *nomen* – name; *calare* – call) means the naming of chemical compounds.

▶ SYSTEMATIC AND TRADITIONAL NOMENCLATURE

The names of organic compounds may be systematic or traditional (trivial). The International Union of Pure and Applied Chemists (IUPAC) publish definitive rules for the nomenclature of inorganic and organic compounds, extracts from which can be found in the *Handbook of Chemistry and Physics*, published by the Chemical Rubber Co. in the USA. Most examination boards in the UK use the very much shorter set of rules, *Chemical Nomenclature, Symbols and Terminology*, published by the Association for Science Education (ASE). The ASE publication includes a list of both recommended systematic and traditional names. Examiners prefer recommended systematic names, especially for organic compounds, but a correct traditional name is better than an incorrect recommended name!

The systematic names of inorganic compounds of an element capable of existing in several oxidation states will follow the Stock notation and include the oxidation number – for instance: $KMnO_4$ potassium manganate(VII), K_2MnO_4 potassium manganate(VI) and $MnCl_2$ manganese(II) chloride. The systematic name of an organic compound conveys information about the structure and usually consists of prefix(es), root and suffix together with numbers and punctuation. For example:

prefix	2-methyl	CH_3	
root	propan-	$CH_2-\overset{\mid}{C}H-CH_3$	2-methylpropan-2-ol
suffix	2-ol	$\underset{\mid}{O}H$	

The prefix(es) and root are often derived from the names of the alkanes. The prefix(es) and suffix usually describe the **functional groups** present in the molecular structure.

◀ Alkanes, Functional group, Oxidation number, Transition metals ▶

NUCLEAR MAGNETIC RESONANCE

Nuclear magnetic resonance (NMR) spectroscopy is a technique for determining molecular structure by analysing the absorption of radio-frequency electromagnetic radiation by the sample substance when it is suspended in a very strong magnetic field. The sample substance is dissolved in tetrachloromethane (CCl_4), containing a small amount of tetramethylsilane $Si(CH_3)_4$ as a reference substance. The solution is placed in a tiny glass sample tube, which is placed between two electrical coils at right-angles to each other and at the centre of an extremely powerful electromagnet. One coil transmits radio-frequency waves from an oscillator and the other acts as a detector. At an appropriate frequency and magnetic field strength, the sample substance absorbs energy from the transmitter coil and induces a signal in the detector coil. An NMR spectrum may be produced by increasing the magnetic field strength and keeping constant the frequency of the transmitted radio-waves, or by increasing the frequency of the transmitted radio-waves and keeping constant the magnetic field strength.

The absorption of energy by the sample substance is dependent upon the magnetic properties of ^1H, ^{13}C, ^{14}N, ^{19}F and ^{31}P nuclei present in its molecular structure. The nuclear magnetic properties of these atoms differ from one another; they depend upon their molecular environment. As a result, NMR spectra can be used to detect and distinguish between hydrogen atoms present in OH, CH_2, and CH_3 groups in a molecule of , say, ethanol (Fig. N.3).

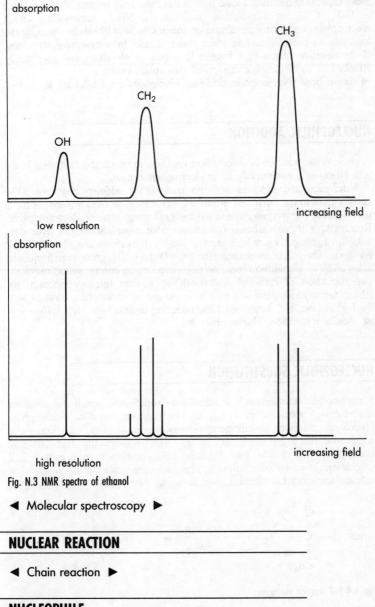

Fig. N.3 NMR spectra of ethanol

◄ Molecular spectroscopy ►

NUCLEAR REACTION

◄ Chain reaction ►

NUCLEOPHILE

A nucleophile is an electron-pair donor and a **Lewis base**. The nucleophilic atom capable of donating a **lone pair** of electrons may be part of an **ion** (such as OH^-, CN^-) or a **molecule** (such as H_2, NH_3). Nucleophiles attack **electrophiles** and form **coordinate** or **dative covalent bonds** by donating the lone pair to the electrophilic atom under attack. In a molecular structure, electronegative atoms that have a lone pair of electrons and are readily attacked by an electrophile are called nucleophilic centres.

◀ Lewis base, Nucleophilic addition, Nucleophilic substitution ▶

NUCLEOPHILIC ADDITION

A nucleophilic addition is an **addition reaction** in which the first step is the attachment of a **nucleophile** to an electrophilic centre.

Aldehydes and **ketones** undergo nucleophilic addition reactions. The carbon atom in the **carbonyl group** $>C=O$ acts as an electrophilic centre in part because the oxygen atom in the carbonyl group is more electronegative. Reactions of these carbonyl compounds with **aqueous** sodium hydrogen-sulphite, hydrogen cyanide, ammonia and its derivatives are all nucleophilic additions. The reactions involving the $>C=O$ and $-NH_2$ groups are frequently followed by an **elimination reaction** to give an overall **condensation reaction**. The **reduction** of carbonyl compounds by sodium tetrahydridoborate or lithium tetrahydridoaluminate may be regarded as nucleophilic addition with the hydride ion, H^-, supplied by the **reducing agents** (see Fig. A.16).

◀ Addition reaction, Nucleophile ▶

NUCLEOPHILIC SUBSTITUTION

A nucleophilic substitution is a **substitution reaction** in which the attacking (and leaving) atom (or group) is an electron-pair donor or **nucleophile**. **Hydrolysis** reactions of **halogenoalkanes** are nucleophilic substitutions in which the more powerful nucleophile OH^- replaces the less powerful nucleophile, the halide ion. Primary halogenoalkanes favour the S_N2 mechanism in which substitution by the nucleophile takes place in one slow rate-determining **bimolecular step** involving two species (see Fig. N.4).

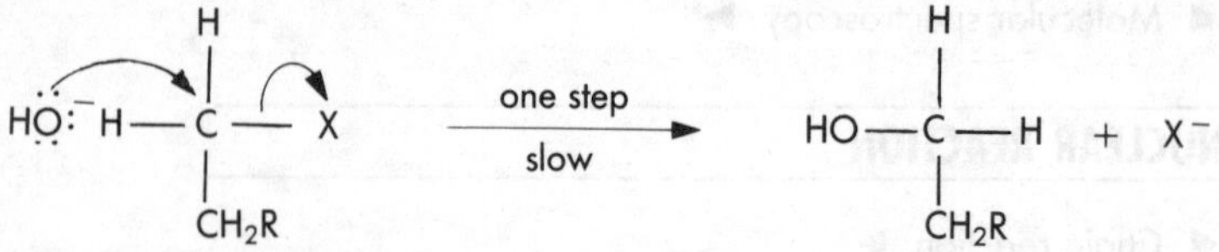

Fig. N.4 S_N2 reaction mechanism

Tertiary halogenoalkanes favour the S_N1 mechanism in which substitution by the nucleophile takes place in one slow rate-determining unimolecular step involving one species followed by a second fast step (see Fig. N.5).

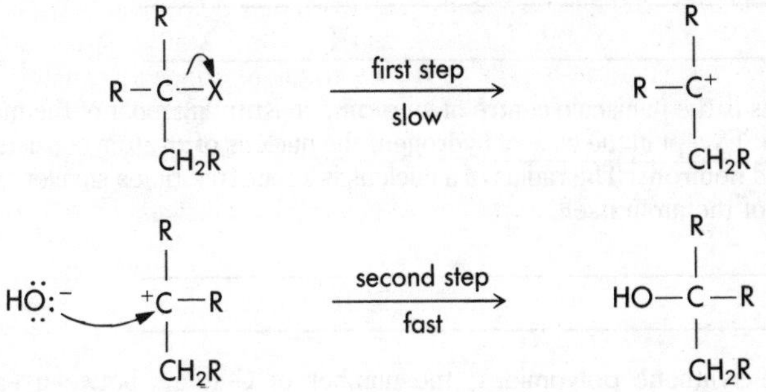

Fig. N.5 S_N1 reaction mechanism

Substitution reactions of secondary halogenoalkanes can proceed by either or both mechanisms. Nucleophilic substitutions of halogenoalkanes can be important for putting some functional groups into a molecule – see Fig. N.6, which shows substitution reactions for a typical halogenoalkane, RCH_2Br.

Nucleophile	Reagents and conditions	Substitution products
$HO:^-$	sodium hydroxide aqueous; heat under reflux	RCH_2OH alcohol
$CH_3O:^-$	sodium methoxide $(CH_3ONa)^*$ in methanol; under reflux	RCH_2OCH_3 ether
$:CN:^-$	potassium cyanide (KCN) in ethanol; under reflux	RCH_2CN nitrile
$AgCN:$	silver cyanide (AgCN) in ethanol; under reflux	RCH_2NC isonitrile
$:NH_3$	ammonia (NH_3) aqueous or gaseous; heat in sealed tube under pressure	RCH_2NH_2 amine – (primary)
$:NH_2CH_2R$		$(RCH_2)_2NH$ – (secondary)
$:NH(CH_2R)_2$	with excess RCH_2Br	$(RCH_2)_3N$ – (tertiary)
$:N(CH_2R)_3$		$(RCH_2)_4N^+$ quaternary ammonium bromide

* made by adding sodium metal to methanol: $Na + CH_3OH \rightarrow CH_3ONa + \frac{1}{2}H_2$

Fig. N.6 Substitution reactions of a typcial halogenoalkane

Acylation with an **acyl chloride** may be regarded as substitution of the chlorine atom by nucleophiles such as **alcohols**, ammonia and **amines**.

$$ROH + CH_3COCl \rightarrow CH_3CO_2R + HCl$$

$$NH_3 + CH_3COCl \rightarrow CH_3CONH_2 + HCl$$

$$RNH_2 + CH_3COCl \rightarrow CH_3CONHR + HCl$$

◀ Nucleophile, Reaction mechanism, Substitution ▶

NUCLEUS

The nucleus is the miniscule centre of an **atom**, constituting most of the mass of the atom. Except in the case of hydrogen, the nucleus of an atom consists of **protons** and **neutrons**. The radius of a nucleus is about 10^{12} times smaller than the radius of the atom itself.

NYLON

Nylons are synthetic **polyamides**; the number of C-atoms between each amide linkage may vary from one nylon to another. The first nylon was produced in 1930 by Dr W H Carothers, working for Du Pont, and is thought to have been named after the cities of New York and LONdon.

Two of the most important nylons are nylon-6 and nylon-6,6. The numbers indicate the number of carbon atoms in a molecule of each **monomer**. Nylon-6 is derived from just one monomer, 6-aminohexanoic acid, whereas nylon-6,6 is derived from two monomers: 1,6-diaminohexane and hexane-1,6-dioic acid.

$$-(CH_2)_5-\underset{O}{\overset{\|}{C}}-\underset{H}{\overset{|}{N}}-(CH_2)_5-\underset{O}{\overset{\|}{C}}-\underset{H}{\overset{|}{N}}- \qquad -\underset{H}{\overset{|}{N}}-(CH_2)_6-\underset{H}{\overset{|}{N}}-\underset{O}{\overset{\|}{C}}-(CH_2)_4-\underset{O}{\overset{\|}{C}}-\underset{H}{\overset{|}{N}}-$$

$$\text{nylon-6} \qquad\qquad\qquad \text{nylon-6,6}$$

These long-chain **condensation polymers** are extremely tough, with **hydrogen-bonding** between CO and NH groups in neighbouring polymer chains contributing considerably to the strength of the nylon. A major use of nylon is in the manufacture of cord which is at least twice as strong as steel and which is used to strengthen car tyres. Other uses of nylon include low-friction moving parts for machinery, and fibres for textiles.

OCTET RULE

When **atoms** combine, they have a tendency to achieve a **noble gas electronic configuration**.

OIL

Oils are viscous liquids which are usually immiscible with water. *Mineral* oils such as **petroleum** are mixtures of **hydrocarbons**. *Natural* oils obtained from animals and plants are either the so-called *essential* oils, consisting of mixtures of simple **esters** and **unsaturated hydrocarbons** based on 2-methylbutadiene (isoprene) units, or the **triglycerides** derived from unsaturated long-chain **fatty acids** and propane-1,2,3-triol.

Oils, unlike fats, are liquid at about 20°C and the hydrocarbon chain of the aliphatic acids is unsaturated; for example:

$CH_3(CH_2)_7CH=CH(CH_2)_7CO_2CH_2$

$CH_3(CH_2)_7CH=CH(CH_2)_7CO_2CH$

$CH_3(CH_2)_7CH=CH(CH_2)_7CO_2CH_2$

olein; an ester of octadec-9-enoic (oleic) acid and propane-1,2,3-triol (glycerol)

Oils, like fats, can be hydrolysed by boiling with concentrated aqueous alkali to produce glycerine and soaps. Oils can also be hydrogenated or 'hardened' to form fats in the production of margarine.

◀ **Fat, Hydrogenation, Triglycerides, Soaps** ▶

OPTICAL ACTIVITY

Optical activity is the ability of substances to rotate the plane of plane-polarised light. It is shown by inorganic and organic compounds. Inorganic compounds (for instance, calcium carbonate as calcite) that do not crystallise in the cubic system form **crystals** that are **anisotropic** to **polarised light** but they usually lose their optical activity in solution. Some inorganic **complexes** form stereoisomeric **ions** that retain their optical activity in solution – see Fig. O.1.

Optical activity is found widely in **organic chemistry** and occurs with compounds whose **molecules** do not have a plane of symmetry. The commonest assymetric molecules are those with a chiral centre resulting from four different **groups** attached to the same carbon **atom**. Such molecules can exist in two different forms or **isomers** that are non-superimposable mirror images of each other. These isomers are called optical isomers or **enantiomers** and their existence is known as optical isomerism or enantiomorphism. Enantiomers have identical physical properties, except that one isomer will be **dextrorotatory** and the other will be **laevorotatory**. The isomers also have identical chemical properties except for their reaction with another chiral molecule.

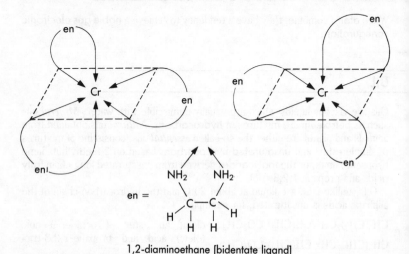

1,2-diaminoethane [bidentate ligand]

Fig. 0.1 Stereoisomerism in complex ions

Apart from glycine, **amino acids** can exist as enantiomers because the α-carbon atom is the chiral centre or assymetric carbon atom (see Fig. A.29). Note that the L-form of an α-amino-acid may be (+) dextro- or (−) laevo-rotatory. An equimolar mixture of enantiomers such as L(−)glutamic acid and D(−)glutamic acid is called a **racemic mixture** or a racemate. These mixtures are not optically active because the rotations of the plane of the polarised light by the dextrorotatory and laevorotaatory enantiomers are equal, but in opposite directions, and cancel. Racemic mixtures are usually obtained during the **synthesis** of potentially optically active compounds because the two enantiomers usually have an equal chance of being formed in a reaction. Molecules with two identical chiral centres can have a third isomer, called a *mesomer*, which is not optically active, because the optical activity of one chiral centre internally compensates for the optical activity of the other chiral centre:

D-enantiomer   L-enantiomer   mesomer

2,3-dihydroxybutane-1,4-dioic (tartaric) acid

◀ Dextrorotatory, Enantiomer, Laevorotatory, Stereoisomerism ▶

OPTICAL ISOMERISM

◀ Optical activity ▶

OPTICAL ROTATION

Optical rotation is the change in the orientation of the plane of plane-polarised light caused when the light passes through an anisotropic crystal or a solution of an enantiomer of an optically active compound. The angle of rotation of the plane of plane-polarised light is measured with a polarimeter. A positive rotation is a rotation in the clockwise direction as viewed towards the source of light; a negative rotation is anticlockwise. Dextrorotatory substances produce a positive rotation and are labelled (+) while laevorotatory substances produce a negative rotation and are labelled (−), as in D(+)glucose and D(−)fructose, for example.

◀ Optical activity, Polarimeter ▶

ORBITAL

An orbital is a mathematical probability distribution function describing the location of an electron in an atom or molecule. The term orbital is used for the region of space occupied by one or two (but no more than two) electronic charge clouds. The probability of finding an electron at any point is usually represented by the density of shading in a diagram (see Fig. O.2). Different orbitals have different shapes. The s-orbital is radially symmetrical about the nucleus and the three p-orbitals are dumb-bell shapes lying along axes mutually at right-angles. The five d-orbitals and the seven f-orbitals have directional character.

It is acceptable to say that an electron *occupies* or *is in* an orbital. Electrons in an atom have kinetic energy because they move and potential energy because they repel one another and are attracted by the nucleus. It is quite

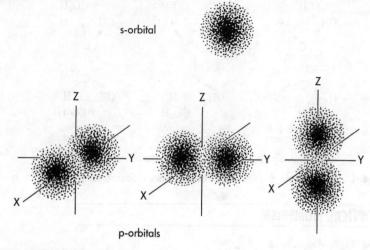

s-orbital

p-orbitals

Fig. 0.2 Electron orbitals

usual to describe the energy of an electron occupying an orbital as the *energy of an orbital* and to speak of different orbitals having different energies.

◀ Electronic configuration ▶

ORDER OF A REACTION

The order of a reaction is the sum of the powers (m and n) of the concentrations in the mathematical equation expressing the **rate of the reaction** in terms of the **concentrations** of the substances affecting the rate.

In the general equation, rate = $k[A]^m[B]^n$ m and n are the orders of reaction with respect to reactant A and reactant B, respectively, and $m + n$ is the overall order of the reaction. For the kinds of reaction studied in A-level courses, the orders may be 0, 1 or 2 but $m + n$ will not exceed 2. For example, the reaction of iodine with propanone in aqueous acid is **zero-order** with respect to iodine, **first-order** with respect to propanone, first-order with respect to hydrogen ion and **second-order** overall:

$$CH_3COCH_3(aq) + I_2(aq) \xrightarrow[\text{catalyst}]{\text{acid}} CH_3COCH_2I(aq) + H^+(aq) + I^-(aq)$$

The **hydrolysis** (or inversion) of sucrose to a mixture of fructose and glucose is first-order with respect to sucrose and pseudo-first-order overall, because there is so much water present that its concentration hardly changes:

$$C_{12}H_{22}O_{11}(aq) + H_2O(l) \rightarrow CH_2OH(CHOH)_3COCH_2OH(aq) +$$
$$CH_2OH(CHOH)_4CHO(aq)$$

▶ FINDING THE ORDER OF A REACTION

There are two ways of finding the order of a reaction. One method is to determine how the concentration of each substance affects the initial rate. The initial rate is measured because at the very start of the reaction the concentrations of all the substances will be known. The second method, known as the Ostwald isolation method, is to determine how the concentration of each substance affects the rate when the concentrations of all the other substances are so high that their values barely change during the reaction.

◀ Rate of reaction, Reaction mechanism ▶

ORGANIC CHEMISTRY

Organic chemistry is the study of carbon compounds. The word organic (Greek *organon* – part of the body) was first used in 1807 by Jöns Jacob Berzelius, Professor of Medicine and Pharmacy at Stockholm. Until Friedrich Wohler synthesised urea in 1828, it was believed that organic compounds could be produced only by living organisms. Since then more than 6 million organic compounds have been identified and named, but the number is increasing daily. The compounds are divided into classes according to their molecular structure and their properties are interpreted in terms of the chemistry of their functional groups.

◀ Functional group ▶

ORGANOMETALLIC COMPOUND

An organometallic compound is a compound in which a metal atom is attached directly to a carbon atom in an organic group. Organometallic chemistry is a large and important branch of chemistry. Important examples for A-level chemistry include aluminium trialkyls, used in the production of stereoregular polymers, and Grignard compounds, widely used in the synthesis of organic compounds.

◀ Grignard compound, Ziegler–Natta catalysts ▶

OSMOSIS

Osmosis is the movement of water or some other solvent through a semipermeable membrane from a solution of high concentration to a solution of lower concentration.

OSMOTIC PRESSURE

The osmotic pressure of a solution is the minimum pressure required to prevent osmosis into the solution separated from pure solvent by a semipermeable membrane.

The osmotic pressure (π) of a solution is a colligative property given by the expression $\pi V = nRT$ where R is the **gas constant** and n is the amount of solute in a volume of solution V at thermodynamic (absolute) temperature T.

Osmometry is the technique of measuring osmotic pressures with an osmometer. It is particularly important as a method of determining **molar masses** of natural and **synthetic macromolecules**.

◀ Colligative property, Ideal gas equation ▶

OXIDATION

Oxidation is a process involving the gain of oxygen or similar electronegative non-metals, loss of hydrogen or loss of electrons. The process leads to an increase in the **oxidation number** of an element.

OXIDATION NUMBER

The oxidation number is a number assigned to an element according to the following set of rules, which are applied in a priority order.

	Rule	**oxidation number**
1	Oxidation number of an uncombined element	0
2	Sum of oxidation numbers of elements in uncharged formula	0
3	Sum of oxidation numbers of elements in charged formula	charge
4	Oxidation number of fluorine in any formula	−1
5	Oxidation number of an alkali metal in any formula	+1
6	Oxidation number of an alkaline earth metal in any formula	+2
7	Oxidation number of oxygen (except in peroxides = −1)	−2
8	Oxidation number of halogen in metal halides	−1
9	Oxidation number of hydrogen (except in metal hydrides = −1)	+1

These rules are founded on the principle that we imagine **electrons** to transfer completely to the more electronegative **atom** when two atoms combine and then we define the oxidation number for each element as the number of electrons to be added to each combined atom to make it uncharged. Since fluorine is the most electronegative of all the elements and has seven valence electrons, we always imagine a combined F atom to have one negative charge and therefore imagine the subtraction of one electron to make a combined F atom uncharged. Hence the oxidation number of combined fluorine is always −1.

The maximum oxidation numbers of the first twenty elements exhibit **periodicity** (see Fig. O.3). The **p-block** and **s-block** metals show little or no variety in oxidation number, whereas the **d-block transition metals** show a wide range of oxidation numbers (see Fig. O.4).

Positive oxidation numbers are used in the Stock system of naming inorganic compounds of d-block and p-block elements with variable valence in

oxidation
numbers

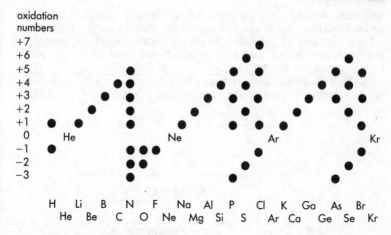

Fig. 0.3 Periodicity of oxidation numbers

order to specify unambiguously their composition. The oxidation number of the element is written as a Roman numeral in brackets immediately after the part of the name containing the element – for instance, copper(II) dichromate(VI) $CuCr_2O_7$ and tetracarbonylnickel(O) $Ni(CO)_4$.

Oxidation numbers can help in balancing equations for **redox reactions**.

electronic configurations	Sc	Ti	V	Cr	Mn	Fe	Co	Ni	Cu	Zn
4s	[↑↓]	[↑↓]	[↑↓]	[↑]	[↑↓]	[↑↓]	[↑↓]	[↑↓]	[↑]	[↑↓]
3d	[ ]	[ ]	[ ]	[↑]	[↑]	[↑]	[↑]	[↑]	[↑↓]	[↑↓]
	[ ]	[ ]	[ ]	[↑]	[↑]	[↑]	[↑]	[↑]	[↑↓]	[↑↓]
	[ ]	[ ]	[↑]	[↑]	[↑]	[↑]	[↑]	[↑↓]	[↑↓]	[↑↓]
	[ ]	[↑]	[↑]	[↑]	[↑]	[↑]	[↑↓]	[↑↓]	[↑↓]	[↑↓]
	[↑]	[↑]	[↑]	[↑]	[↑]	[↑↓]	[↑↓]	[↑↓]	[↑↓]	[↑↓]

oxidation states	Sc	Ti	V	Cr	Mn	Fe	Co	Ni	Cu	Zn
					[7]					
				6	6	6				
			[5]	5	5	5	5			
		[4]	4	4	[4]	4	4	4		
	[3]	3	3	[3]	3	[3]	[3]	3	3	
		2	2	2	[2]	[2]	[2]	[2]	[2]	[2]
		1	1	1	1	1	1	1	[1]	

Fig. 0.4 Oxidation states of the transition metals

OXIDES

An oxide is a binary compound of an element with oxygen. Oxides with the

oxidation number of oxygen $= -2$ are usually classified as acidic, amphoteric, basic and neutral. Oxides with the oxidation number of oxygen $= -1$ are classified as peroxides. In general, acidic oxides are formed by non-metals, amphoteric oxides by metalloids and basic oxides by metals, thereby giving rise to a broad pattern in the periodic table (Fig. O.5). Basic oxides have **giant structures** which are often ionic, whereas acidic oxides have simple molecular structures which are covalent (see Fig. O.6). The **s-block**

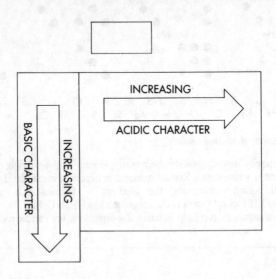

Fig. O.5 Trend in acid–base character of oxides

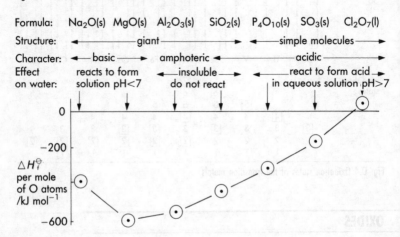

Fig. O.6 Trend in properties of oxides

Group 1	Group 2
normal oxides (O^{2-})	normal oxides (O^{2-})
$4Li(s) + O_2(g) \rightarrow 2Li_2O(s)$ $4Na(s) + O_2(g) \rightarrow 2Na_2O(s)$	$2Be(s) + O_2(g) \rightarrow 2BeO(s)$ $2Mg(s) + O_2(g) \rightarrow 2MgO(s)$ $2Ca(s) + O_2(g) \rightarrow 2CaO(s)$
sodium forms a mixture of the normal oxide and the peroxide on burning in air	$2Sr(s) + O_2(g) \rightarrow 2SrO(s)$ $2Ba(s) + O_2(g) \rightarrow 2BaO(s)$
peroxides ($^-O\!-\!O^-$)	peroxides ($^-O\!-\!O^-$)
$2Na(s) + O_2(g) \rightarrow Na_2O_2(s)$	$Ba(s) + O_2(g) \rightarrow BaO_2(s)$
superoxides (O_2^-)	superoxides (O_2^-)
$K(s) + O_2(g) \rightarrow KO_2(s)$ $Rb(s) + O_2(g) \rightarrow RbO_2(s)$ $Cs(s) + O_2(g) \rightarrow CsO_2(s)$	there are stable superoxides of Na, Ca, Sr and Ca but they are not formed by burning in air

Fig. O.7 s-block oxides

oxidation number										
+7					Mn_2O_7					
+6				CrO_3						
+5			V_2O_5							
+4		TiO_2	VO_2	CrO_2	MnO_2					
+3	Sc_2O_3	Ti_2O_3	V_2O_3	Cr_2O_3	Mn_2O_3	Fe_2O_3				
+2		TiO	VO		MnO	FeO	CoO	NiO	CuO	ZnO
+1									Cu_2O	

Fig. O.8 d-block oxides

metals (see Fig. O.7) burn vigorously in air to form normal oxides, peroxides or superoxides depending upon their reactivity. The **d-block metals** form a variety of oxides (see Fig. O.8).

The oxidising power of the oxides becomes greater with increasing oxidation number of the metal. The oxides in the highest oxidation state are covalent and acidic, while the rest are largely ionic and neutral, amphoteric or basic depending upon the oxidation number of the metal, (see Fig. C.21). Those with an oxidation number of + 2 are mostly basic.

The **p-block elements** such as B, Al, Si, Ge, As and Sb which are close to the diagonal dividing line of the **periodic table** tend to form amphoteric oxides. Elements on the right of the p-block form acidic oxides many of which are often called **acid anydrides**. Carbon and nitrogen also form the neutral oxides CO, NO and N_2O. The oxides of the **halogens** are generally explosive liquids, but iodine(V) oxide atypically forms stable white **crystals**. The structure and properties of the trioxides and pentoxides of nitrogen and phosphorus should be compared – see Fig. O.9.

N₂O₃(g) nitrogen trioxide

unstable simple molecule decomposes into NO and NO₂ so structure could be:

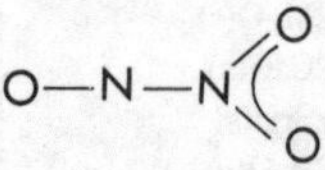

$$O-N-N\begin{array}{c}O\\\\O\end{array}$$

anhydride of nitrous acid:
$N_2O_3(g) + H_2O(l) \rightarrow 2HNO_2(aq)$

N₂O₅(s) nitrogen pentoxide

unstable simple molecule decomposes into O₂ and NO₂:

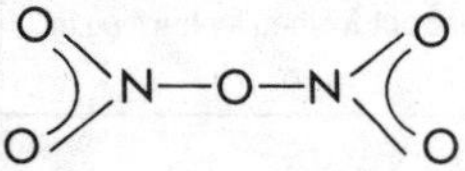

$$\begin{array}{c}O\\\\O\end{array}N-O-N\begin{array}{c}O\\\\O\end{array}$$

anhydride of nitric acid:
$N_2O_5(g) + H_2O(l) \rightarrow 2HNO_3(aq)$

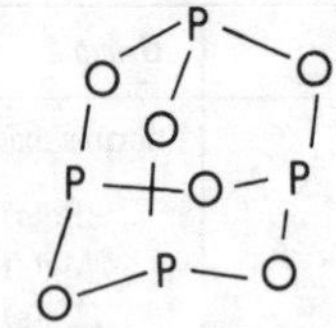

P₄O₆(s) phosphorus(III) oxide

phosphorus atoms at corners of a tetrahedron joined by bonds to oxygen atoms

anhydride of phosphoric acid:
$P_4O_6(s) + 6H_2O(l) \rightarrow 4H_3PO_3(aq)$

P₄O₁₀(s) phosphorus(V) oxide

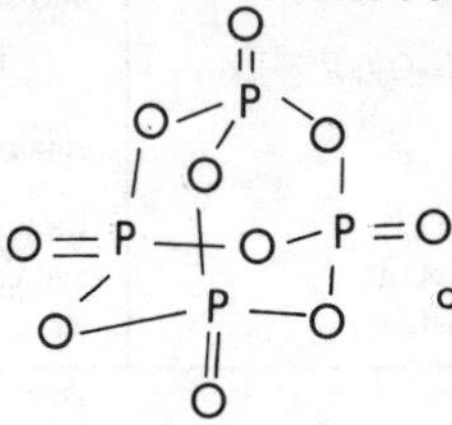

polymeric solid but molecule of vapour has similar structure to trioxide molecule but with another oxygen atom attached to phosphorus atoms forming tetrahedra

anhydride of phosphoric acid:
$P_4O_{10}(s) + 6H_2O(l) \rightarrow 4H_3PO_4(aq)$

Fig. 0.9 Comparing the oxides of nitrogen and phosphorus

OXIDISING AGENT

An oxidising agent or *oxidant* is a substance which causes the **oxidation** of another substance, known as a **reducing agent** or *reductant*.

OXOACID

An oxoacid is an acid which donates a **proton** from an OH group in its molecular structure; its resulting conjugate base is an oxoanion. The mineral acids nitric HNO_3 or $HONO_2$ and sulphuric H_2SO_4 or $(HO)_2SO_2$ are typical monoprotic and diprotic oxoacids respectively.

PAPER CHROMATOGRAPHY

Paper chromatography is a technique to separate a mixture of solutes using liquids as the mobile phase and paper as the support medium and stationary phase.

A tiny spot of a solution of the mixture (M) is placed near the edge of the paper. This edge of the paper is then kept in touch with a suitable solvent which by capillary action gradually seeps past the spot towards the opposite edge of the paper. The solvent carries the different solutes in the mixture different distances along the paper, because some adsorb on to the paper more strongly than others and because some are more soluble in the water supported by the paper than in the moving solvent. Just before the solvent reaches the opposite edge, the paper is taken out of contact with the solvent and dried. The position of the solute spots are located by some suitable means and the chromatogram analysed. Coloured components can usually be seen with the naked eye. Some colourless substances may be visible in ultraviolet

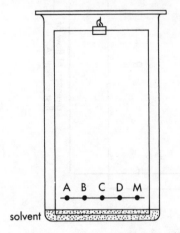
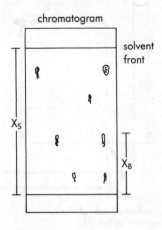

A–D: spots of known substances
M: spot of mixture containing A,B,C
Rf value of B $= X_B \div X_S$

Fig. P.1 One-way paper chromatography

light; others must be made visible by spraying the paper with chemicals which react with the solutes to give coloured substances. For example, **amino acids** give mauve spots when the chromatogram is sprayed with ninhydrin and kept at about 110°C for ten minutes.

In *one-way chromatography* (Fig. P.1) spots of known solutes (A, B, C, D) may be placed on the paper for comparison. In two-way chromatography, a spot of the mixture for analysis is placed near the corner of the paper. Chromatography is carried out in one direction using a suitable solvent and then, after the paper is dried, it is carried out in another direction at right-angles, using a second suitable solvent to separate the components the first solvent failed to separate. The separated solutes may be identified by measuring their R_f values – the ratio of the distance travelled by the solute to the distance travelled by the solvent front (see Fig. P.2).

◀ Chromatography, Thin-layer chromatography ▶

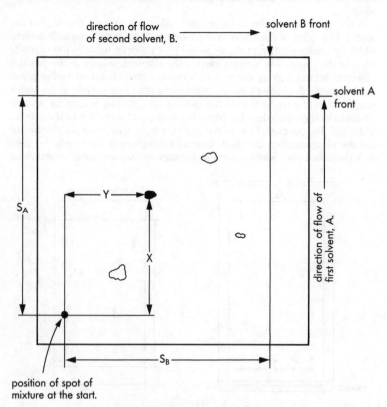

Rf value in solvent A is $X \div S_A$
Rf value in solvent B is $Y \div S_B$

Fig. P.2 Two-way paper chromatogram

PARAFFINS

The word paraffins (Latin *para* – little; *affinis* –reactivity) is the traditional, unsystematic and almost obsolete word for **alkanes**.

PARENT MOLECULE ION

The parent molecule ion is the **cation** formed in a **mass spectrometer** when a **molecule** does not fragment but just loses one **electron**. The parent molecule ion is often – but not always – detected as the highest mass-to-charge ratio peak in a **mass spectrum**.
◄ Mass spectrometer ►

PARTIAL PRESSURE

In a mixture of two or more gases, the partial pressure of each gaseous component is the pressure which that gas would exert if it alone occupied the volume taken up by the mixture of gases.
◄ Dalton's law of partial pressures ►

PARTIAL VAPOUR PRESSURE

The partial vapour pressure is the partial pressure of the vapour in a mixture of vapours in **equilibrium** with a mixture of two or more volatile substances.
◄ Vapour pressure ►

p-BLOCK ELEMENTS

The p-block elements (Fig. P.3) consist of the **halogens**, the **noble gases** and the elements of Groups III to VI inclusive. These groups of elements have little in common; their physical and chemical behaviour changes across the block from the metals on the lower left to the non-metals on the upper right, via the metalloids dividing the block diagonally from top left to bottom right. These trends *across* the block are shown clearly by the Group IV elements and their compounds. Similarities in a group are shown clearly by the halogens and the noble gases.

Certain individual p-block elements need to be studied in detail for A-level chemistry; these are shown in Fig. P.4.
◄ Halogens, Noble gases ►

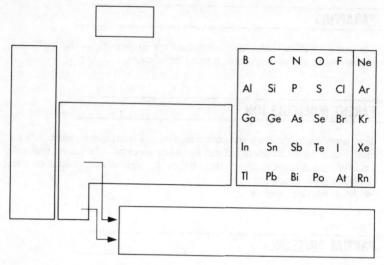

Fig. P.3

B (He)$2s^2 2p^1$	C $2s^2 2p^2$	N $2s^2 2p^3$	O $2s^2 2p^4$	F $2s^2 2p^5$	Ne $2s^2 2p^6$
Al (Ne)$3s^2 3p^1$	Si $3s^2 3p^2$	P $3s^2 3p^3$	S $3s^2 3p^4$	Cl $3s^2 3p^5$	Ar $3s^2 3p^6$
Ga (Ar)$3d^{10}4s^2 4p^1$	Ge $4s^2 4p^2$	As $4s^2 4p^3$	Se $4s^2 4p^4$	Br $4s^2 4p^5$	Kr $4s^2 4p^6$
In (Kr)$4d^{10}5s^2 5p^1$	Sn $5s^2 5p^2$	Sb $5s^2 5p^3$	Te $5s^2 5p^4$	I $5s^2 5p^5$	Xe $5s^2 5p^6$
Tl (Xe)$4f^{14}5d^{10}6s^2 6p^1$	Pb $6s^2 6p^2$	Bi $6s^2 6p^3$	Po $6s^2 6p^4$	At $6s^2 6p^5$	Rn $6s^2 6p^6$

He $1s^2$

Fig. P.4 Individual p-block elements to be studied in detail

PEPTIDES

A peptide is an organic compound composed of two or more **amino acids** joined together by a peptide linkage (−CO−NH−).

A *dipeptide* consists of two amino acids – for instance, NH$_2$−CHR−CO−NH −CHR−CO$_2$H, where R is a side-group such as CH$_3$. A **polypeptide** may contain up to about 300 amino acid units. Polypeptides are produced by partial **hydrolysis** of **proteins**.

◀ Amide, Polyamide, Protein ▶

PERIOD

A period is a horizontal row of elements in the **periodic table**. Hydrogen and helium constitute the first short period. The elements from lithium to neon and from sodium to argon make up the second and third short periods. The fourth period, from potassium to krypton, includes the first transition series, from scandium to zinc. The fourth, fifth, sixth and seventh periods are called the long periods. From left to right across a period the properties of the elements change from metallic to non-metallic.

◄ Group, Periodic table ►

PERIODICITY

Periodicity is a term which refers to the recurrence of similar chemical and physical properties at regular intervals of atomic number when elements are arranged in increasing **atomic number** order.

In the periodic table the regular intervals are counted in elements, but the size of the interval is NOT constant. The number of elements in each interval changes from 2 to 32 in the sequence: 2, 8, 8, 18, 18, 32. The recurrence of similar properties is exemplified by the Group I **alkali metals**:

| **alkali metal** | (hydrogen) | lithium | sodium | potassium | rubidium | caesium | francium |

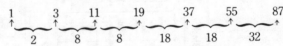

| **atomic no.** | 1 | 3 | 11 | 19 | 37 | 55 | 87 |
| **interval** | | 2 | 8 | 8 | 18 | 18 | 32 |

◄ Alkali metals, Chlorides, Halogens, Hydrides, Oxides ►

PERIODIC LAW

The properties of the elements are a periodic function of **atomic number**.

Mendeleyev's statement of this law appeared in 1869, referring to chemical properties: 'The elements, if arranged according to their atomic weights, exhibit an evident periodicity of properties. Elements which are similar as regards their chemical properties have atomic weights which increase regularly.' In 1870 Julius Lothar Meyer published his famous curve of **atomic volume** plotted against atomic weight and defined the periodic law as follows: 'The properties of the elements are largely periodic functions of the atomic weight.' In the modern version of the law, atomic weight is replaced by atomic number.

◄ Atomic volume, Atomic weight, Periodicity ►

PERIODIC TABLE

The periodic table is an arrangement of all the elements in increasing **atomic**

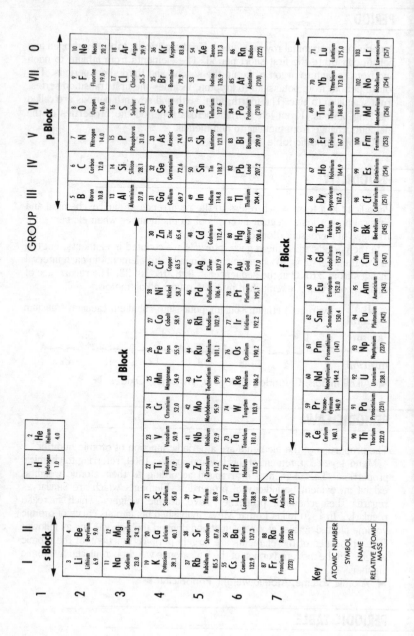

Fig. P.5

number order to form four areas, known as the s-block, p-block, d-block and f-block (Fig. P.5).

The s- and p-block elements are known as the *main-group* elements. The s-block elements are divided vertically into the **alkali metals** (Group I) and the **alkaline earth metals** (Group II). The **p-block elements** are divided vertically into six groups, which include the **halogens** (Group VII) and the **noble gases** (Group O). In each group the elements have similar properties and their atoms have the same outer-shell electronic structure. Down each group the elements show a gradation of properties with increasing atomic number. Across a period from the alkali metals to the noble gases, the properties of the main-group elements change from metallic to non-metallic.

The **d-block elements** are divided horizontally into three transition series of metals with characteristic properties. The first series, from scandium to zinc, includes the transition metals, which form an important part of the inorganic chemistry of every A-level syllabus. The f-block elements are divided horizontally into two rows of fourteen almost identical elements called the **lanthanoids** (lanthanides) and **actinoids** (actinides).

Credit for the existence of the modern form of the periodic table is usually given to the work of Dmitri Ivanovic Mendeleyev, a Russian professor of chemistry at St Petersburg. In 1869 he published his table and his statement of the **periodic law**. An important feature of Mendeleyev's work was his successful prediction of the existence and properties of gallium, scandium and germanium – three elements named after the countries where these rare minerals were discovered.

The periodic table is a fundamentally important means of organising **inorganic chemistry**. The broad patterns in the physical and chemical properties of the elements (and their compounds) down groups and across a period should be related to the **electronic configurations** of the elements. An illustrated periodic table, in the form of a large wallchart published by Time Life International Books is available from the Royal Society of Chemistry, (Burlington House, Piccadilly, London W1V 0BN). The table is in colour, with detailed notes and photographs of each element. Copies can be ordered through bookshops or directly from the Distribution Centre, Blackhorse Road, Letchworth, Hertfordshire, SG6 1HN.

PERSPEX

Perspex is the trade name for a form of the addition **polymer** known systematically as poly(2-methylpropenoate) and traditionally as poly(methylmethacrylate).

◀ Addition polymerisation, Polymer ▶

PETROCHEMICAL

Petrochemicals or petroleum chemicals are organic chemicals produced on an industrial scale from **petroleum** and natural gas. Methane, ethene, propene, butene and butadiene are the most important raw materials obtained in this way. They serve as starting compounds for a very wide range of petrochemical compounds.

PETROLEUM

Petroleum (Latin *petra* – rock; *oleum* – oil) is a naturally occurring mixture of **hydrocarbons**, commonly called crude oil. It is fractionally distilled, and some of the fractions are thermally and catalytically cracked, to provide the raw materials for the petrochemical industry.

◀ Cracking, Distillation, Fractional distillation ▶

pH

$$pH = -\log_{10}([H_3O^+(aq)]/mol\ dm^{-3})$$

The definition of pH proposed in 1909 by S.P. Sørensen, a Danish biochemist, provides a scale of acidity which usually ranges from about ($[H_3O^+(aq)] = 1 \times 10^{-1}\ mol\ dm^{-3}$) to 14 ($[H_3O^+(aq)] = 1 \times 10^{-14}\ mol\ dm^{-3}$). At 25°C a neutral solution has a pH = 7 because in pure water at 25°C $[H_3O^+(aq)] = [OH^-(aq)] = 1 \times 10^{-7}\ mol\ dm^{-3}$.

◀ Hydrogen ion, Neutral solution, pH meter ▶

PHENOLS

Phenols are a class of **aromatic** compounds with one or more –OH groups attached directly to a benzene ring.

▶ NOMENCLATURE

Phenol is the simplest member, with only one –OH group attached to an unsubstituted benzene ring. Names of substituted phenols contain the name of the **substituent group** and its position on the benzene ring (see Fig. P.6).

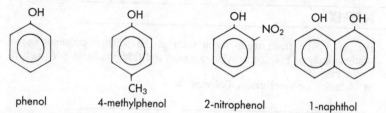

phenol   4-methylphenol   2-nitrophenol   1-naphthol

When the –OH group is attached to two fused benzene rings the compound is called a naphthol.

Fig. P.6 Naming phenols

► *PROPERTIES*

Phenol is a white crystalline solid. It is very toxic and a strong skin irritant. Impurities or sunlight turn the crystals pink or red. The solid absorbs moisture from the air but is only sparingly soluble in water. In contrast to ethanol, phenol (old name *carbolic acid*) is a **weak acid** ($pK_a = 9.9$) which reacts with **aqueous** sodium hydroxide to give a solution of sodium phenate – see Fig. P.7. The ability of the phenol **molecule** to donate a **proton** is explained by the $C_6H_5O^+$ anion being stabilised by **delocalisation** of its negative charge (see Fig. P.8).

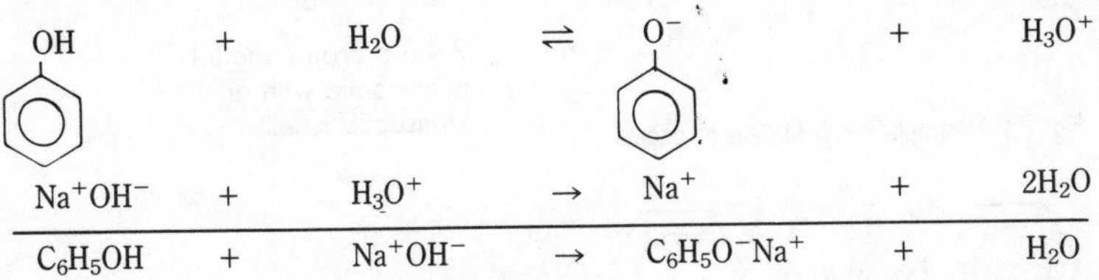

OH	+	H_2O	$\rightleftharpoons$	O^-	+	H_3O^+
Na^+OH^-	+	H_3O^+	$\rightarrow$	Na^+	+	$2H_2O$
C_6H_5OH	+	Na^+OH^-	$\rightarrow$	$C_6H_5O^-Na^+$	+	H_2O

Fig. P.7 Phenols dissolve in aqueous alkali

negative charge on the oxygen atom
becomes part of the delocalised π-
electrons of the benzene ring and
helps to stabilise the phenoxide ion

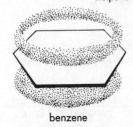

benzene   phenoxide ion

Fig. P.8 Delocalised structure of the phenoxide anion

In contrast to ethanoic acid, phenol is weaker than carbonic acid (H_2CO_3), so it does NOT react with **carbonates** and hydrogencarbonates to give off carbon dioxide. This important difference between phenol and carboxylic acids is often tested in A-level examinations.

Phenol turns 'neutral' aqueous iron(III) chloride from yellow-orange to purple. The phenolate ions $C_6H_5O^-$ are strong **ligands** which displace water molecules from the hexaaquairon(III) cations, $[Fe(H_2O)_6]^{3+}$(aq), to form a new **complex** such as $[Fe(H_2O)_4(C_6H_5O)_2]^+$(aq). This diagnostic test often appears in exams and practical work involving **qualitative analysis**.

Phenol undergoes rapid **electrophilic substitutions** (Fig. P.9) in the ortho/para (2,4,6) positions because the $-OH$ group activates the benzene ring. So, when phenol is shaken with aqueous bromine, the orange-brown solution is decolorised and a white precipitate forms.

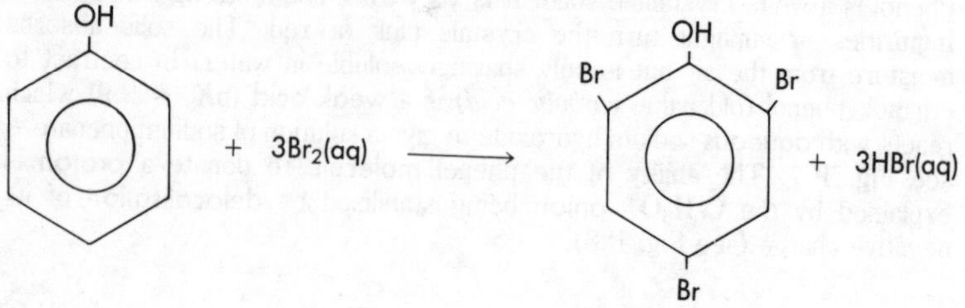

2,4,6-tribromophenol
(white solid with an
'antiseptic' smell)

Fig. P.9 Electrophilic substitution of phenol

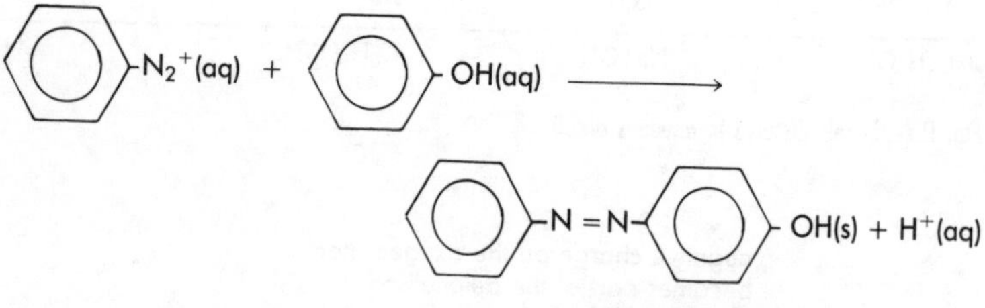

4-hydroxyazobenzene
an orange solid

Fig. P.10 Nucleophilic phenols couple with electrophilic diazonium ion

Phenols and naphthols in aqueous sodium hydroxide undergo nucleophilic coupling with ice-cold aqueous electrophilic diazonium ions to form strongly coloured precipitates of azo dyes (see Fig. P.10).

Phenol is manufactured by the 'cumene' process in which propanone is a valuable co-product. Phenol is used in the production of epoxy-, phenolic and poly(carbonate) resins and of caprolactam – an intermediate in the manufacture of nylon.

◀ Alcohol, Hydroxyl group ▶

PHENYL GROUP

The phenyl group, $C_6H_5^-$, is a benzene molecule with one hydrogen atom missing.

pH METER

A pH meter is an instrument for measuring the pH of a solution by determining the e.m.f. of an electrochemical cell, one-half of which incorporates the solution whose pH is being measured. The complete pH meter usually consists of a combination glass/Ag–AgCl reference electrode connected to a high-resistance voltmeter (often wrongly called the pH meter). The 'electrode' is really an electrochemical cell with the solution missing from one of its half cells. The 'electrode' is dipped into the test solution to complete the half-cell and the pH read from the e.m.f. scale (calibrated in pH units) – see Fig. P.11.

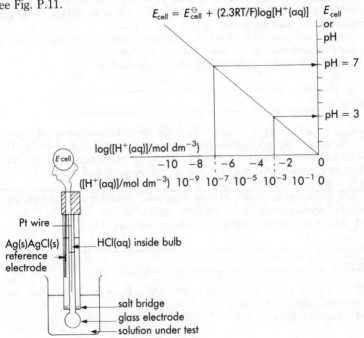

Fig. P.11 Measuring pH with an electrochemical cell

The connection between the e.m.f. of the cell (E_{cell}^{θ}) and pH ($-\log[H^+(aq)]$) is given by the **Nernst equation**: $E_{cell} = E_{cell}^{\theta} + (2.3RT/F)\log[H^+(aq)]$.

PHYSICAL CHEMISTRY

Physical chemistry is the study of the structure of substances and the kinetics and thermodynamics of changes in substances. At A-level, physical chemistry deals with such topics as atomic structure, bonding, gas laws, changes of state, energy changes, chemical equilibria, electrochemistry, rates of reaction and catalysis. The theories developed in physical chemistry provide a basis for understanding the facts, patterns and principles of **inorganic** and **organic** chemistry. Physico-chemical theories are usually quantitative and often mathematically expressed.

◀ **Inorganic chemistry, Organic chemistry** ▶

pK_a

$pK_a = -\log_{10}(K_a/\text{mol dm}^{-3})$ where K_a is the **dissociation constant** of an acid.

Although in principle pK_a values may be less than 0 or greater than 14, they are normally recorded in the range from 1 to 14. pK_a for ethanoic acid is 4.8 and is typical of the **weak carboxylic acids**. The pK_a of an acid and the pK_b of its **conjugate base** are related: $pK_a + pK_b = pK_w$.

pK_a for any weak acid = pH of a maximum buffering solution containing equal **concentrations** of the weak acid and its conjugate base.

◀ Acid dissociation constant, Equilibrium constant ▶

pK_b

$pK_b = \log_{10}(K_b/\text{mol dm}^{-3})$ where K_b is the **dissociation constant** of a base.

pK_b values, like pK_a values, normally lie in the range from 1 to 14. The value for aqueous ammonia is 4.8 and is similar for **aliphatic amines**. pK_b for a base may be calculated from the pK_a of its **conjugate acid** and pK_w:
$pK_b = pK_w - pK_a$.

pK_b for any weak base = pOH of a maximum buffering solution containing equal **concentrations** of the weak base and its conjugate acid.

◀ Base dissociation constant, Equilibrium constant ▶

pK_w

$pK_w = \log_{10}(K_w/\text{mol}^2 \text{ dm}^{-6}) = 14$ at 25°C where $K_w = [H_3O^+(aq)][OH^-(aq)]$ (the ionic product for water).

◀ Neutral solution, pH ▶

PLANE POLARISED LIGHT

Plane polarised light is electromagnetic radiation whose electric vector is in one plane only. It is usually produced by passing ordinary visible light through a polaroid filter or a nicol prism. The nicol prism is made from a crystal of Iceland spar (a form of **calcite**) by cleaving it into two pieces and cementing the pieces together at very precise angles, using Canada balsam.

◀ Optical activity ▶

PLASTICS

Plastics are **synthetic polymers** which can be moulded or formed into any shape.

Thermosetting plastics are usually three-dimensional cross-linked **condensation polymers**, typified by the phenol-formaldehyde (phenol-methanal) Bakelite resins produced in the USA in 1909 by the Belgian-born chemist, Leo Hendrik Baekeland. Once formed into shape by cross-linking polymerisation, thermosetting polymers are hard, rigid plastics that cannot be remoulded by heating.

Thermoplastic polymers are typified by the **addition polymers** formed by the **alkenes**. These plastics are generally softer than the thermosetting plastics. Thermoplastics can be softened by heating and then remoulded.

◀ Thermoplastic polymer, Thermosetting polymer ▶

POLAR BOND

A polar **covalent bond** is formed when two **atoms** having different **electronegativities** share a pair of **electrons** unequally.

In a molecule of hydrogen chloride, the shared pair of electrons is attracted more strongly by the chlorine **nucleus** than by the hydrogen nucleus. Consequently, the chlorine has more than its fair share of the bonded electron pair, both atoms are slightly (δ) charged and the hydrogen chloride molecule ($\delta + H - Cl \delta -$) is polar. The greater the difference between the electronegativity index (N_p) values of the two atoms, the greater the polarity of the bond. Polar bonds (such as C–Cl) result in a polar molecule (such as CH_2Cl_2) with a permanent **dipole moment** when they reinforce one another and in a non-polar molecule (for instance CCl_4) when they cancel each other out. The dipole moment (p) of a **diatomic molecule** is a measure of the polarity of the bond between the two atoms.

◀ Dipole–dipole attraction, Intermediate bond ▶

POLARIMETER

A polarimeter (Fig. P.12) is an instrument for measuring the angle of rotation of the plane of **plane polarised light** by solutions of **optically active** substances.

◀ Optical acitivity ▶

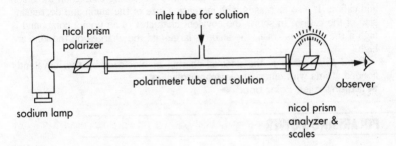

inlet tube for solution

nicol prism
polarizer

polarimeter tube and solution

observer

sodium lamp

nicol prism
analyzer &
scales

Fig. P.12

POLARISABILITY

Polarisability refers to the ease with which the shape of an anion can be distorted by the attraction of a **cation** for its outer-shell electrons. The polarisability of an anion increases as its charge increases and as its radius increases. An alkaline earth metal **carbonate** decomposes into an **oxide** on heating in a bunsen flame – for example, $MgCO_3(s) \rightarrow MgO(s) + CO_2(g)$. The thermal stability of oxide compared to the thermal instability of the carbonate can be interpreted in terms of the relative polarisabilities of the large CO_3^{2-} anion and the smaller O^{2-} anion.

◀ Polarising power ▶

POLARISATION

Anion polarisation is the distortion of the shape of a polarisable **anion** caused by the attraction of a polarising **cation**. When cations and anions form a **crystal lattice**, the net positive charge on the cations attracts the outer-shell **electrons** of the anions. This may polarise the anion (distort its shape) and cause the compound to be less ionic (more covalent) in character – see Fig. P.13.

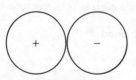
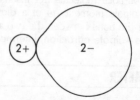

ionic bonding with no
covalent character

small cation polarises large anion
to give some covalent character

Fig. P.13 Polarisation of anions

The polarisation of an anion increases with increasing charge on the anion and cation. It also increases with increasing size of the anion and decreasing size of the cation. In short, the covalent character of an (ionic) compound is high if the cation is small, the anion is large and the charges on the ions are high.

Bond polarisation is the unequal sharing of **electrons** by two covalently bonded atoms with different **electronegativities**.

◀ Fajan's rules, Polar bond ▶

POLARISING POWER

Polarising power is the ability of a **cation** to distort the shape of an **anion** by attracting its outer-shell electrons. The polarising power of a cation increases as its charge increases and as the **ionic radius** decreases – that is, as its *charge density* increases. The increasing covalent character of the **chlorides** NaCl, $MgCl_2$, $AlCl_3$ and $SiCl_4$ from NaCl (ionic) to $SiCl_4$ (covalent) reflects the increase in the polarising power of the cation.

◀ Fajan's rules, Polarisability ▶

POLAR MOLECULE

A polar molecule is a molecule with a permanent **dipole moment**. A **diatomic molecule** (such as $H-Cl$) has a permanent **dipole moment** because the more electronegative atom (chlorine) polarises the **covalent bond** (with hydrogen) by taking an unequal share of the electron pair. Polyatomic molecules (such as CH_2Cl_2 and CCl_4) possessing **polar bonds** (such as $C-Cl$) will have a permanent dipole moment if the polar bonds reinforce one another, but they will have no permanent dipole moment if the polar bonds cancel each other out:

permanent dipole

non-polar molecule

◀ Intermediate bond ▶

POLAR SOLVENT

A polar solvent is a liquid consisting of **polar molecules**. Liquid ammonia, the lower **alcohols**, methanoic acid and liquid hydrogen fluoride are all polar solvents, but water is the best known and most widely used.

Many ionic compounds dissolve in polar solvents because the polar molecules are attracted by the ions, which become *solvated* and disperse into the solvent. The solvent molecules cause the electrostatic forces of attraction between the ions to diminish as they penetrate the **crystal lattice**. The **solvation** energy given out compensates for the **lattice energy** taken in. The small difference between these two large energy terms is the energy of the solution process. Some polar covalent compounds dissolve and ionise in polar solvents. For example, HCl(g) and $CH_3CO_2H(l)$ dissolve well in water, the former **strong acid** ionising completely and the latter **weak acid** ionising partially:

$$HCl(g) + H_2O(l) \rightarrow H_3O^+(aq) + Cl^-(aq)$$

$$CH_3CO_2H(l) + H_2O(l) \rightleftharpoons CH_3CO_2^-(aq) + H_3O^+(aq)$$

POLYAMIDE

Polyamides are **condensation polymers** in which the **monomers** are held together in a long chain by the amide linkage $-CO-NH-$. The two most important classes of polyamide are the synthetic **nylons** and the naturally occurring **proteins**.

◀ Condensation polymer, Nylon, Polypeptide, Protein ▶

POLYESTER

A polyester is a **condensation polymer** derived from a polyhydric alcohol and a polycarboxylic acid. The most common linear polyesters are derived from benzene-1,4-dicarboxylic acid (terephthalic acid) and ethane-1,2-diol and are used to make polyester fibres under such trade names as 'Terylene' and 'Dacron'. Polyester (alkyd) resins for use in paints are derived from linear polyesters crosslinked by including propane-1,2,3-triol in the esterification process (see Fig. P.14).

◀ Condensation polymer, Ester ▶

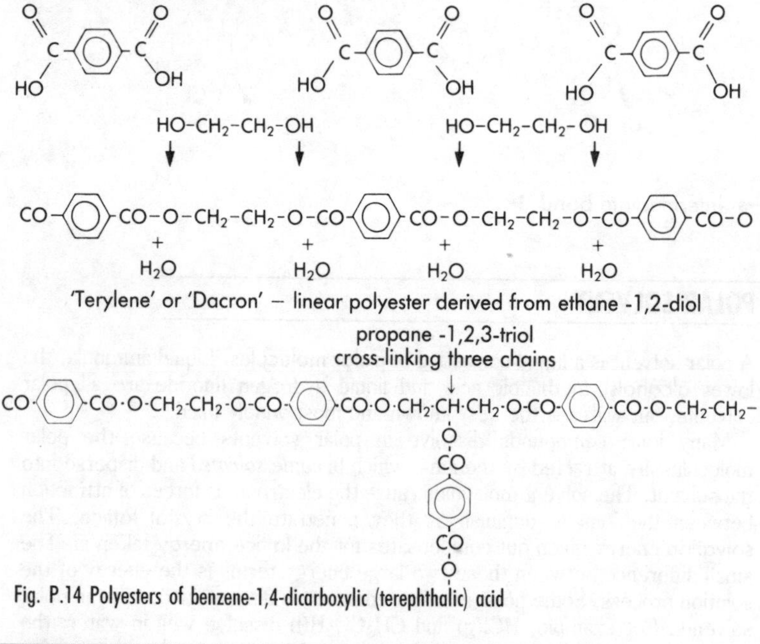

'Terylene' or 'Dacron' – linear polyester derived from ethane-1,2-diol

propane-1,2,3-triol
cross-linking three chains

Fig. P.14 Polyesters of benzene-1,4-dicarboxylic (terephthalic) acid

POLYMER

A polymer is a natural or synthetic **macromolecule** formed by the combination of a large number of small molecules called **monomers**. Polysaccharides and proteins are examples of natural polymers. Nylons, poly(alkenes) and polyesters are examples of synthetic polymers.

POLYMERISATION

Polymerisation is a reaction in which a large number of small molecules (monomers) combine to form a small number of large molecules (polymers).

ADDITION POLYMERISATION

Addition polymerisation is the combination of **unsaturated monomers** (usually **alkenes**) to form **saturated polymers** (usually **alkanes**):

$$n{>}C{=}C{<} \rightarrow (-C-C-)_n$$

Low-density poly(ethene) 'plastic' film for bags and food packaging is manufactured on the industrial scale quite cheaply using high temperature and very high pressure (for instance, 200°C and 1,500 atm) **chain reactions**, initiated by **radicals** (for instance, trace of oxygen), to produce branched-chain stereo-irregular **(atactic)** polymers. High-density poly-(ethene) for rigid containers like milk crates is manufactured by a more expensive process using a **Ziegler–Natta catalyst** of titanium(IV) chloride with an **organo-metallic aluminium compound** (such as triethylaluminium, $Al(C_2H_5)_3$) in an inert solvent at a lower temperature and pressure (60°C and 2 atm, for example) to produce stereo-regular **(isotactic and syndiotactic)** polymers.

CONDENSATION POLYMERISATION

Condensation polymerisation is the formation of condensation polymers, such as **polyesters** and **polyamides**, by the combining of molecules into larger molecules with the eliminating of small molecules:

$$n-OCO-(CH_2)_5-NH_3^+ \rightarrow -CO-(CH_2)_5-NH\left[-CO-(CH_2)_5-NH\right]_{(n-2)} -CO$$

nylon-6 salt nylon-6

$$-(CH_2)_5-NH-CO- + (n-1)H_2O$$

water

Most condensation polymerisation reactions are a kind of co-polymerisation involving two different monomers. For example, polyesters are derived from a dicarboxylic acid and a dihydric alcohol.

POLYMORPHISM

Polymorphism is a term which refers to substances which can exists in two or more different forms (called *polymorphs*) in the same physical state. The polymorphs of a substance have different physical and/or chemical properties. Allotropy is the special case of polymorphism in which the substances are elements. In many cases the polymorphs of a substance consist of the same chemical units arranged in different **crystal lattices**. For example, **calcite**, chalk, limestone and marble are some of the polymorphs of calcium carbonate, all of which consist of calcium ions and carbonate ions.

◀ Allotropy/allotrope ▶

POLYPEPTIDE

A polypeptide is an organic compound composed of many **amino acids** joined together by a peptide linkage ($-CO-NH-$). Polypeptides may contain around 100 to 300 amino acid units. They are obtained when proteins are partially hydrolysed.

◀ Amino acid, Peptide, Protein ▶

POLYSACCHARIDES

A polysaccharide is a carbohydrate consisting of a long chain of monosaccharide units. Cellulose and starch are important polysaccharides which are derived from glucose.

◀ Carbohydrate ▶

PRIMARY ALCOHOL

A primary **alcohol** is an alcohol containing the group $-CH_2OH$. Primary alcohols differ from secondary and tertiary alcohols by forming aldehydes when partially oxidised and carboxylic acids when completely oxidised. Secondary alcohols form **ketones** and tertiary alcohols resist oxidation by reagents such as acidified aqueous potassium dichromate(VI).

◀ Alcohol, Hydroxyl group ▶

PROMOTER

A promoter is a substance which increases the activity of a **catalyst**. Aluminium oxide promotes the activity of metallic iron used as the **heterogeneous catalyst** in the **Haber process**.

◀ Catalyst ▶

PROPAGATION

The propagation stage in a **radical chain reaction** is the second stage in which **free radicals**, from the initiation stage, participate in a chain of reactions that form the products and regenerate the free radicals.

In the radical chain reactions of gaseous chlorine or bromine with alkanes or hydrogen, alkyl radicals and hydrogen atoms are formed as intermediates that act as chain carriers; for example –

$$Cl\cdot + CH_4 \rightarrow HCl + CH_3\cdot \text{ then } CH_3\cdot + Cl_2 \rightarrow CH_3Cl + Cl\cdot$$

$$Br\cdot + H_2 \rightarrow HBr + H\cdot \text{ then } H\cdot + Br_2 \rightarrow HBr + Br\cdot$$

The net effect of the propagation steps is the overall change of reactants into products – $Cl_2 + CH_4 \rightarrow HCl + CH_3Cl$ and $Br_2 + H_2 \rightarrow 2HBr$.

◀ Radical chain reaction ▶

PROTEIN

Proteins are naturally occurring **polyamides**, derived from no more than about twenty-five different **amino acids**, having **molar masses** ranging from about 5×10^3 to 4×10^7 g mol^{-1}.

Proteins can be hydrolysed, by hot **concentrated aqueous acid** or alkali, at the amide linkages to produce a mixture of **peptides** and, ultimately, α-amino acids which can be analysed by **chromatography**.

α-amino acid

The sequence of α-amino acids in the protein chains constitutes the *primary structure* of the protein. $-S-S-$ bonds, **hydrogen bonding** and electrostatic attractions involving side groups can cause chains to cross-link, coil into a helix and fold into pleats to produce the *secondary structure* of the protein. The *tertiary structure* is the overall three-dimensional shape of the cross-linked, coiled and pleated protein molecule. For the globular proteins such as **enzymes**, antibodies, haemoglobin, casein, albumin and insulin, the tertiary structure gives a compact spherical molecule which is soluble in water. For the fibrous proteins such as keratin and collagen in hair and muscle, the tertiary structure gives molecules consisting of long coiled strands or flat sheets which are insoluble in water. Enzymes are an important group of biochemical **catalysts**.

◀ Enzyme ▶

PROTON

A proton (p) is a sub-atomic particle with a positive charge of approximately

1.602×10^{-19} C and a mass of 1.673×10^{-27} kg. The mass of a proton is just slightly smaller than the mass of a **neutron**. Protons together with neutrons are called *nucleons*. They constitute the **nucleus** and most of the mass of an atom. The number of protons in the nucleus of an atom is called the **atomic number** or **proton number** and is the same for all the **isotopes** of a particular element.

When the nucleus of a radioisotope emits an α-particle, the proton number falls by 2, and when it emits a β-particle the proton number rises by 1. In each case the isotope of a different element is formed. Free protons rarely exist. In aqueous solution the proton is datively bonded to a water molecule as the oxonium ion H_3O^+ which itself is hydrated. $H_3O^+(aq)$ is frequently written as $H^+(aq)$ for simplicity. In **acid—base** reactions, protons are donated by acids and accepted by bases.

◀ Atomic number, Moseley's experiments ▶

PROTON ACCEPTOR

A proton acceptor is a **base**.
◀ Bronsted—Lowry theory ▶

PROTON DONOR

A proton donor is an **acid**.
◀ Bronsted—Lowry theory ▶

PROTON NUMBER

The proton number is a term sometimes used instead of **atomic number** (Z), the number of protons in a **nucleus**.

PROTON TRANSFER

A proton transfer reaction is an **acid—base reaction** in which an acid, the proton donor, donates or transfers a **hydrogen ion** H^+ to a base, the proton acceptor.
◀ Bronsted—Lowry theory ▶

QUALITATIVE ANALYSIS

Qualitative analysis is the branch of analytical chemistry concerned with establishing the identity and nature of chemical substances. In A-level practical chemistry, mixtures may be separated and the components identified using standard techniques such as **paper chromatography**, and simple test-tube reactions involving changes in colour, formation of precipitates and production of gases. More advanced instrumental techniques include **gas chromatography, infra-red absorption spectroscopy, mass spectrometry, nuclear magnetic resonance** and **X-ray crystallography**. These techniques can be used for qualitative analysis but they are also included under the heading of **quantitative analysis** because they can make measurements on composition and structure.

QUANTITATIVE ANALYSIS

Quantitative analysis is the branch of analytical chemistry concerned with measurements on chemical substances. In A-level practical chemistry, quantitative analysis may include standard techniques involving the **titration** of solutions, the measurement of gas volumes and the weighing of precipitates. More advanced quantitative analysis employs instrumental techniques such as **gas chromatography, infra-red absorption spectroscopy, mass spectrometry, nuclear magnetic resonance** and **X-ray crystallography**.

QUANTUM NUMBER

A quantum number is one of four numbers which arise from the quantum-mechanical wave theory of the atom and define the properties of an extra-nuclear **electron** in an **atom**.

The *principal* quantum number (n) gives the main energy level of an electron and can have the values 1, 2, 3, and so on. This quantum number is shown as an integer in the **electronic configuration** of atoms and ions. The *orbital* (or *azimuthal*) quantum number (I) defines the subshells and can have the values 0 (s-subshell), 1 (p-subshell), 2 (d-subshell) and 3 (f-subshell). This quantum number is shown by the letters s, p, d and f in the electronic configuration of atoms and ions. The *magnetic* quantum number (m) governs the number of orbitals in a subshell by taking integer values from $+1$ to -1 so that there is only one s-orbital but there are three p-orbitals, five d-orbitals and seven f-orbitals. The *spin* quantum number can have the values $+\frac{1}{2}$ or $-\frac{1}{2}$.

According to the exclusion principle proposed by Wolfgang Pauli in 1925, no two electrons in an atom may have the same four quantum numbers. Consequently an orbital has only two electrons with opposite spin quantum numbers ($+\frac{1}{2}$ and $-\frac{1}{2}$) often represented by [↑↓].

QUATERNARY AMMONIUM ION

A quaternary ammonium ion is a tetrahedral cation consisting of a nitrogen atom to which is attached four alkyl groups – for instance, $(CH_3)_4N^+$.

RACEMIC MIXTURE

A racemic mixture (±) is an equimolar mixture of the **dextrorotatory** (+) and **laevorotatory** (−) forms of **optically active** substance. In 1848, Louis Pasteur applied the name racemic acid (Latin *racemus* − a bunch of grapes) to an optically inactive dicarboxylic acid formed during wine-making. In attempting to crystallise the ammonium sodium salt of the acid, Pasteur achieved the rare spontaneous **resolution** of the racemate into distinguishable separate **crystals** of dextro- and laevorotatory ammonium sodium tartrate. Normally **crystallisation** of an aqueous racemate produces only identical crystals of the racemate with the dextro- and laevorotatory forms contained in the same **crysal lattice**. Racemic mixtures are usually obtained during the synthesis of potentially optically active compounds because the two **enantiomers** usually have an equal chance of being formed in a reaction.

Racemic mixtures are not optically active because the rotations of the plane of the polarised light by the dextrorotatory and laevorotatory enantiomers are equal but in opposite directions and cancel:

D-enantiomer   L-enantiomer

2,3-dihydroxybutane-1, 4-dioic (tartaric) acid

RADICAL

Radical is an almost obsolete term for two or more **atoms** combined together as a **group** which confers its distinctive and characteristic properties upon the compound containing it.

In many cases the term radical was used synonymously with **ion**. For example, the ammonium (NH_4) and nitrate (NO_3) radicals correspond to the ammonium (NH_4^+) and nitrate (NO_3^-) ions. The atoms in a radical or in a compound ion tend to remain bound together as a group and behave chemically as a single unit.

◀ Free radical ▶

RADICAL CHAIN REACTION

A radical chain reaction is a gas-phase reaction in which **free radicals** come into existence in an **initiation** step, act as chain carriers in the **propagation** steps and go out of existence in **termination** steps.

◀ Chain reaction ▶

RADIOACTIVITY

Radioactivity is the spontaneous disintegration of atomic nuclei accompanied by the emission of **alpha particles**, **beta particles** and gamma radiation.

RADIUS RATIO

The radius ratio is the radius of the smaller **ion** divided by the radius of the larger ion in a **crystal lattice**. The ratio is usually (but not always) related to the **coordination number** of the smaller ion as follows:

radius (small ion) / radius (large ion)	< 0.41	> 0.41 and < 0.73	> 0.73
coordination number	4	6	8

In a metallic crystal lattice of identical ions, the radius ratio is 1 and the coordination number is either 12 (in the **hexagonal** or **cubic close-packed** structures) or 8 (in the **body-centred structure**).

RAOULT'S LAW

The **partial vapour pressure** (p_A) of component (A) in a solution of miscible volatile liquids is equal to the **vapour pressure** (p_A^o) of the pure component multiplied by its **mole fraction** (x_A) in the solution: $p_A = x_A \times p_A^o$.

This law was proposed by Francois Marie Raoult in 1886. It is obeyed by solutions of substances with closely similar **intermolecular forces**. For example, benzene (b.p. 80°C) and methylbenzene (b.p. 111°C) form a two-component solution that obeys Raoult's law. At 60°C the vapour pressure of pure benzene is about 400 mm Hg and that of pure methylbenzene about 150 mm Hg. According to Raoult's law, the partial vapour pressure of benzene above this solution with methylbenzene will show a linear variation with mole fraction (x_b) from 0 mm Hg (no benzene: $x_b = 0$) up to 400 mm Hg (no methylbenzene: $x_b = 1$). In the same way the partial vapour pressure of methylbenzene will have a linear increase with its mole fraction ($x_m = 1 - x_b$) from 0 to 150 mm Hg. In consequence, the total vapour pressure of the solution will show a linear variation with mole fraction (x_b) from 150 mm Hg (pure methylbenzene) to 400 mm Hg (pure benzene). This information can be displayed in a vapour pressure/composition diagram, as shown in Fig. R.1.

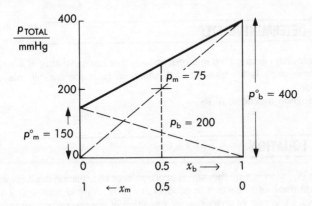

Fig. R.1 Vapour pressure/composition diagram for an ideal solution

Solutions which obey Raoult's law are called *ideal solutions* and are formed by liquids whose molecules have very similar structures and intermolecular forces. *Non-ideal solutions* show positive or negative deviations from Raoult's law because the molecules of the components have different structures and intermolecular forces. Some solutions deviate so much from Raoult's law that they form **azeotropes**.

The majority of non-ideal solutions show positive deviations from Raoult's law. For example, the vapour pressure of a mixture of ethanol and cyclohexane is greater than expected because the **hydrocarbon** molecules come between the ethanol molecules, reducing the **dipole–dipole** and **hydrogen-bonding forces** between them. In contrast, ethyl ethanoate and trichloromethane form a non-ideal solution with a negative deviation from Raoult's law because hydrogen bonding between the ester and the trichloromethane makes the solution less volatile than expected.

RATE CONSTANT

A rate constant (k) is the proportionality constant in a **rate equation**. The value of k does not, of course, depend upon the **concentrations** of the reactants. The units of k are determined by the form of the rate equation and the overall **order** of the reaction. If the rate of a reaction is measured in mol dm^{-3} s^{-1} then the units of k for zero-, first- and second-order reactions are mol dm^{-3} s^{-1}, s^{-1} and mol $^{-1}$ dm^3 s^{-1}, respectively. The rate constant depends upon temperature according to the **Arrhenius equation**. A plot of ln$k$ against $1/T$ (the reciprocal of the reaction temperature in kelvin) gives a straight line with a slope of $-E_a/R$ (where E_a is the **activation energy** and R is the **gas constant**). For many reactions near 298 K, a temperature rise of 10 degrees will approximately double the value of k.

RATE-DETERMINING STEP

The rate-determining step in a **reaction mechanism** consisting of a sequence of simple steps is the slowest step which controls the overall **rate of the reaction**.
◀ Reaction mechanism ▶

RATE EQUATION

A rate equation is a mathematical expression of the relationship between the **concentration** of a reactant or product and time. A *differential* rate equation expresses the rate of a reaction as a function of concentration. An *integrated* rate equation expresses concentration as a function of time. Rate equations can be very complicated, but for A-level chemistry they are restricted to the following general differential rate equations and the associated integrated rate equations:

$$\text{rate} = k[A]^m[B]^n$$

where k is the **rate constant**, m and n are the orders of reaction with respect to reactant A and reactant B, and $m+n$ is the overall **order of the reaction**. For A-level chemistry courses, the orders may be 0, 1 or 2, but $m+n$ will not exceed 2.

RATE OF REACTION

The rate of a reaction is a measure of how quickly the concentrations of reactants decrease with time and the concentrations of products increase with time. The rate is usually expressed as a change of **concentration** (d[ ]) of a specific reactant (or product) in a time interval (dt) so that the rate (d[ ]/dt) is related to the slope of a concentration/time graph and its units are mol dm^{-3} s^{-1} (see Fig. R.2).

The gradients of a graph of reactant concentration against time have negative values. So $-$d[ ]/dt is needed to give positive rate values that get smaller and eventually become zero because reactions slow down and eventually stop, as time goes by.

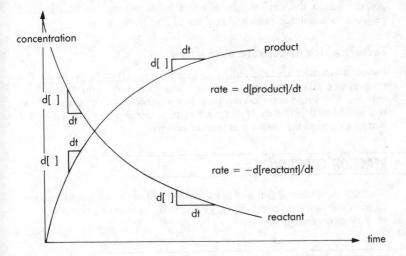

Fig. R.2

▶ MEASURING RATES OF REACTION

The (change in) concentration of a chosen reactant (or product) is determined by observing and measuring a suitable property of the chemical mixture. The techniques for following a reaction include use of a **colorimeter** (for the colour intensity of the mixture), a **polarimeter** (for the angle of rotation of the plane of **polarised light**), a **dilatometer** (for very small changes in volume of the liquid reaction mixture) and an apparatus for measuring the conductance of the mixture.

The reactions studied in an A-level laboratory will proceed at a convenient rate for following by sampling. For example, the concentration of ethanoic acid produced during the **hydrolysis** of ethyl ethanoate can be measured by taking samples of the reaction mixture with a pipette, chilling them in ice-cold distilled water (to slow down the reaction) and titrating with aqueous sodium hydroxide of known concentration.

Instrumental methods based on colorimetry, conductivity, dilatometry or polarimetry are usually better than sampling and titrating the reaction mixture, even when a calibration graph is needed to convert instrument readings into values for the concentration of reactant (or product).

Using a graph

Rates of reaction may be determined from the gradients of a graph of concentration of a reactant (or product) against time. Gradients can be obtained simply (but rather inaccurately) by drawing tangents to a curve, or less easily (but more accurately) by calculation using a computer program. For some chemical reactions (sometimes called 'clock' reactions), a value proportional to the rate may be obtained by measuring the time interval required for a definite amount of reaction to take place.

Factors which affect rate of reaction

Rates of reaction are affected by **catalysts**, concentration, pressure and temperature. The dependence of reaction rate upon concentration is expressed by a **rate equation** involving a **rate constant** and the **orders of the reaction**. The dependence of reaction rate upon temperature is given by the Arrhenius equation, involving **activation energy**.

REACTION COORDINATE

Reaction coordinate or pathway is the name given to the horizontal axis in a diagram showing the change in energy during the course of a reaction.
◀ Activation energy ▶

REACTION MECHANISM

A reaction mechanism is a set of theoretical steps proposed to account for the conversion of reactants into products in an organic chemical reaction. Plausible mechanisms for a reaction are often easy to propose but difficult to substantiate. Experimental evidence for reaction mechanisms is provided by the determination of **rate equations, orders of reaction** and **activation energies**.

For A-level chemistry, a reaction mechanism is usually considered to be a sequence of simple steps. The slowest step is called the **rate-determining step** because it controls the overall **rate of reaction**. The following example is one plausible mechanism for the iodination of propanone:

$$(CH_3)_2CO(aq) + H_3O^+(aq) \xrightarrow{slow} (CH_3)_2COH^+(aq) + H_2O(l)$$

$$(CH_3)_2COH^+(aq) + H_2O(l) \xrightarrow{fast} CH_3-\underset{\underset{OH}{|}}{C}=CH_2 + H_3O^+(aq)$$

$$CH_3-\underset{\underset{OH}{|}}{C}=CH_2 + I_2(aq) \xrightarrow{fast} CH_3-\underset{\underset{O}{\|}}{C}-CH_2I(l) + HI(aq)$$

The first step is **bimolecular** because it involves one propanone molecule and one hydrogen ion as reactants. It is also the rate-determining step because it is the slowest.

The **hydrolysis** of 2-bromo-2-methylpropane by aqueous sodium hydroxide involves a **nucleophilic substitution reaction** in which the rate-determining step is **unimolecular** (S_N1). It is much faster than the S_N2 mechanism for the hydrolysis of 1-bromobutane, in which the rate-determining step is bimolecular (see Fig. R.3).

Bromoalkane RBr	Rate equation	Mechanism
CH_3 | $CH_3 — C — Br$ | CH_3 2-bromo-2-methyl-propane	rate α [RBr]	S_N1 (a two-step mechanism) $RBr \xrightarrow{\text{slow}} R^+$ $R^+ + OH^- \xrightarrow{\text{fast}} ROH$
H | $CH_3CH_2CH_2 — C — Br$ | H 1-bromobutane	rate α [RBr][OH$^-$]	S_N2 (a one-step mechanism) $RBr + OH^- \xrightarrow{\text{slow}} ROH + Br^-$

Fig. R.3 Hydrolysis of bromoalkanes

The meanings of S_N1 and S_N2 should not be confused. '1' stands for unimolecular and means that the rate-determining step involves only one species (one molecule of 2-bromo-2-methylpropane undergoing heterolytic fission). '2' stands for bimolecular and means that the rate-determining step involves two species (one molecule of the electrophilic 1-bromobutane and one nucleophilic hydroxide ion or water molecule). The numbers '1' and '2' do NOT refer to the number of steps in the mechanism.

Organic reactions can be classified into four fundamental types: **addition, elimination, rearrangement** and **substitution**. Their mechanisms can be described in terms of the movement of **electrons** and the **heterolytic** or **homolytic fission** of **polar** and **non-polar bonds** resulting from the attack of **electrophiles, nucleophiles** and **free radicals**

REACTION PATHWAY

◀ Reaction coordinate ▶

REAL GAS

A real gas is a gas whose properties, under certain conditions, do not match those of an ideal gas. Real gases are also called *imperfect* or *non-ideal* gases. Their deviation from ideal gas behaviour increases as the temperature decreases, the pressure increases and the gases approach liquefaction. Under these conditions, the volume of the gas molecules themselves, and the attractive forces between them, become increasingly significant and cannot be ignored. The **van der Waals equation** of state for a real gas attempts to take these two factors into account.

REARRANGEMENT

A rearrangement is a reaction in which the atoms in a molecule or ion of a compound change their relative positions to form a molecule or ion of a different compound. This is one of the four fundamental types of reaction which can account for the great variety of chemical properties of all the various classes of organic compound. Synthetic **organic chemistry** began in 1828 when Friedrich Wohler, while trying to crystallise ammonium cyanate from aqueous solution, obtained urea as a result of an intramolecular rearrangement reaction: $NH_4^+NCO^- \rightarrow (NH_2)_2CO$.

Particular types of rearrangement are named after the chemists who discovered them. An important example of a *Beckmann rearrangement* occurs in the production of nylon-6 with the conversion of cyclohexanone oxime into caprolactam – see Fig. R.4.

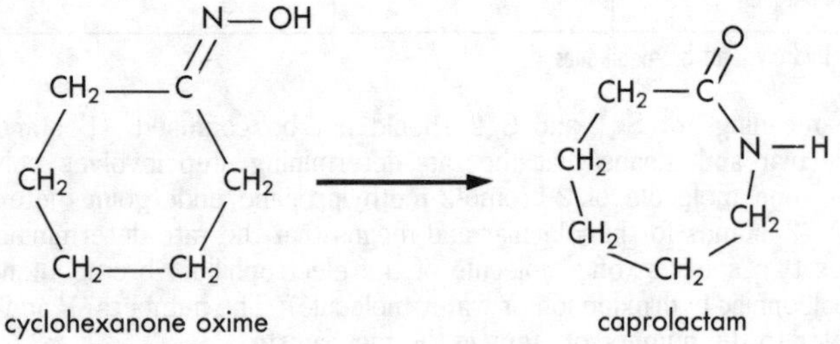

cyclohexanone oxime   caprolactam

Fig. R.4 A Beckmann rearrangement

REDOX REACTION

A redox reaction is a reaction which simultaneously involves **reduction** and **oxidation**. Reduction is a process resulting in a decrease in the **oxidation number** of a combined or uncombined element. Oxidation is a process resulting in an increase in the oxidation number of a combined or uncombined element. These two processes always occur together, and may involve loss or gain of electrons. A redox reaction always involves an increase in oxidation number of one element and a simultaneous decrease in oxidation number of another element, or the same element in the special case of disproportionation.

The changes in the oxidation numbers can help to balance equations for redox reactions – see Fig. R.5. The steps to balance an equation for a redox reaction are:

1 write down the correct formulae of reactants and products
2 assign oxidation numbers using the rules in priority order
3 find the change in oxidation number of the reductant and oxidant
4 balance these two changes in oxidation numbers
5 check and balance if necessary the remaining atoms
6 check the charge balance if there are ions in the equation

step 1 write down correct formulae of all reactants and products:

$$Cr_2O_7^{2-}(aq) + H^+(aq) + SO_3^{2-}(aq) \rightarrow 2Cr^{3+}(aq) + SO_4^{2-}(aq)$$

step 2 apply priority rules to assign oxidation numbers:

$$Cr_2O_7^{2-}(aq) + H^+(aq) + SO_3^{2-}(aq) \rightarrow 2Cr^{3+}(aq) + SO_4^{2-}(aq)$$

$\uparrow$ $\uparrow$ $\uparrow$ $\uparrow$

+6   +4   +3   +6

step 3 calculate rise and fall of oxidation numbers:

(a rise of $6 - 4 = 2$)   +4   +6

              $\downarrow$ $\downarrow$

$$Cr_2O_7^{2-}(aq) + H^+(aq) + SO_3^{2-}(aq) \rightarrow 2Cr^{3+}(aq) + SO_4^{2-}(aq)$$

$\uparrow$ $\uparrow$

$2 \times (+6) = 12$ $2 \times (+3) = 6$ (a fall of $12 - 6 = 6$)

step 4 balance the rise and fall of oxidation numbers:

$$Cr_2O_7^{2-}(aq) + H^+(aq) + 3SO_3^{2-}(aq) \rightarrow 2Cr^{3+}(aq) + 3SO_4^{2-}(aq)$$

                         $\uparrow$

total fall of 6 will equal 3x a rise of 2 to balance Cr and S atoms

step 5 balance other atoms without changing balance of redox atoms:

$$Cr_2O_7^{2-}(aq) + H^+(aq) + 3SO_3^{2-}(aq) \rightarrow 2Cr^{3+}(aq) + 3SO_4^{2-}(aq)$$

$\uparrow$ $\uparrow$ $\uparrow$

seven oxygens   nine oxygens   twelve oxygens

$\Rightarrow 7 + 9 - 12 = $ four oxygens needed on the right hand side; i.e. $4H_2O(l)$

$$Cr_2O_7^{2-}(aq) + H^+(aq) + 3SO_3^{2-}(aq) \rightarrow 2Cr^{3+}(aq) + 3SO_4^{2-}(aq) + 4H_2O(l)$$

          $\uparrow$ $\uparrow$

     one hydrogen   eight hydrogens

$\Rightarrow$ multiply by eight to give fully balanced equation

               $\downarrow$

$$Cr_2O_7^{2-}(aq) + 8H^+(aq) + 3SO_3^{2-}(aq) \rightarrow 2Cr^{3+}(aq) + 3SO_4^{2-}(aq) + 4H_2O(l)$$

step 6 check the charge balance:

$$Cr_2O_7^{2-}(aq) + 8H^+(aq) + 3SO_3^{2-}(aq) \rightarrow 2Cr^{3+}(aq) + 3SO_4^{2-}(aq) + 4H_2O(l)$$

$\uparrow$ $\uparrow$ $\uparrow$ $\uparrow$ $\uparrow$

(2−) $8 \times (1+)$ $3 \times (2-)$ $2 \times (3+)$ $3 \times (2-)$

$\Rightarrow$ (2−) + (8+) + (6−) = (6+) + (6−) $\checkmark$ balanced

Fig. R.5 Balancing equations for redox reactions

In *electrochemistry* all redox reactions are treated as chemical processes involving transfer of **electrons**. Reduction is the gain of electrons by an oxidant, oxidation is the loss of electrons by a reductant, and redox is therefore the transfer of electrons from a reductant to an oxidant:

$$Cl_2(aq) + 2e^- \rightarrow 2Cl^-(aq) \text{ reduction of oxidant}$$
$\uparrow$
oxidant gains electrons

$$Zn(s) \rightarrow Zn^{2+}(aq) + 2e^- \text{ oxidation of reductant}$$
$\uparrow$
reductant loses electrons

$$Zn(s) + Cl_2(aq) \rightarrow Zn^{2+}(aq) + 2Cl^-(aq) \text{ redox reaction}$$
electrons transferred from reductant to oxidant

In an **electrochemical cell**, redox reactions generate a voltage and an electric current. In an **electrolysis** cell, redox reactions are brought about by an applied voltage and the input of an electric current.
◄ Disproportionation ►

REDUCING AGENT

A reducing agent or reductant is a substance which causes the **reduction** of another substance called an **oxidising agent** or oxidant.

REDUCTION

Reduction is a process which involves loss of oxygen or similar electronegative non-metals, gain of hydrogen or gain of electrons, with a consequent decrease in the **oxidation number** of an element.
◄ Redox reaction ►

REFLUXING

Refluxing is the technique of boiling a liquid, condensing the vapour which is given off and running the liquid condensate back into the boiling liquid. The

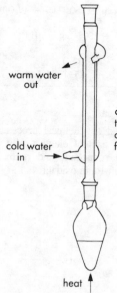

warm water out

cold water in →

condenser set up for refluxing; this means the vapours of the volatile organic compounds are condensed and returned to the flask for continued reaction

heat ↑

Fig. R.6 Laboratory preparation of an aldehyde

technique is widely used in **organic chemistry**. Organic compounds tend to be volatile. Many do not mix well with inorganic reagents, and their reactions are often slow. Refluxing allows organic mixtures to be heated to speed up the reaction without loss of volatile components and helps to keep the reactants well mixed together.

The **oxidation** of **primary alcohols** to **carboxylic acids** is one good example of the use of refluxing (see Fig. R.6).

RELATIVE ATOMIC MASS

The relative atomic mass (A_r) is the ratio of the average mass per **atom** of the natural isotopic composition of an element to one-twelfth of the mass of an atom of nuclide ^{12}C. The actual mass of an atom is so small (the heaviest weight is only about 4×10^{-25} kg) that it is more convenient to compare the masses of atoms to one another and use these relative values.

Since 1808, when John Dalton published his atomic theory, various standards have been used as the basis for **atomic weights**. In 1858 Stanislao Cannizzaro, Professor of Chemistry at Genoa, took the mass of a hydrogen atom as 1 by choosing half the mass of a hydrogen molecule as his standard atomic weight. The Belgian chemist, Jean-Servais Stas (1813–91) proposed that the atomic weight of oxygen should be taken as exactly 16. The discovery of **isotopes** led to the mass of the oxygen isotope ^{16}O being taken as exactly 16. Nowadays the mass of an atom of ^{12}C isotope is taken as exactly 12.000.

The relative atomic mass of an element can be calculated from the natural abundances of its isotopes, determined very accurately from mass spectra. For example, bromine consists of 50.69 per cent ^{79}Br ($A_r = 78.9183$) and 49.31 per cent ^{81}Br ($A_r = 80.9163$), so the relative atomic mass of bromine may be calculated by $(0.5069 \times 78.9183) + (0.4931 \times 80.9163)$ to give $A_r(Br) = 79.9035$. The relative atomic mass of carbon itself is 12.011, because the element consists of two stable isotopes ^{12}C (98.9 per cent) and ^{13}C (1.1 per cent).

Relative atomic mass is a pure number. It does not have units and it should NOT be called 'atomic mass'. Values are listed in data books and may be printed on some versions of the **periodic table**.

RELATIVE MOLECULAR MASS

The relative molecular mass (M_r) is the ratio of the average mass of an entity (having its natural isotopic composition) to one-twelfth of the mass of an atom of nuclide ^{12}C. For example, $M_r(H_2O) = 18.0153$. This term may be used even when the entity is not a molecule. $M_r(NaCl) = 58.44$ can be used instead of 'relative formula mass of sodium chloride' and $M_r(^2H) = 2.0140$ can replace 'relative isotopic mass of deuterium'.

Relative molecular mass is a pure number. It does not have units and it must not be confused with **molar mass**. Values are listed in data books. They

are sometimes wrongly called **molecular weights**. M_r values may be calculated using A_r values and the formula of the entity:
$M_r(NaCl)$ is $A_r(Na) + A_r(Cl) = 22.99 + 35.45$, for example.

RESOLUTION

Resolution is a term which describes the process of separating a racemic mixture into its two **enantiomers** or optically active forms. As a general rule, a racemic mixture cannot be separated by ordinary physical methods such as **fractional crystallisation, distillation** or **chromatography**, because the enantiomers have identical physical properties. An inefficient method is to destroy one of the pair of optical isomers by bacteria or moulds and recover the enantiomer that is not attacked. The principle of an efficient method is to react the racemate with an **optically active isomer** of another compound to produce a mixture of two isomers that have different physical properties because they are not enantiomers. For example, an acidic racemate $(\pm)A$ may be reacted with an optically active base $(-)B$ to give two isomeric salts $(+)A(-)B$ and $(-)A(-)B$ that can be separated by, say, fractional crystallisation and then each individually hydrolysed to recover the optically active base and the separate acidic enantiomer.

The term resolution also refers to the resolving power of instruments in **spectroscopy** and mass spectrometry. A high-resolution **mass spectrometer** may measure the **relative molecular mass** of a **parent molecule ion** to four decimal places and enable the precise **molecular formula** to be determined with the aid of very accurate values for the **relative atomic masses** of the constituent **atoms**.

REVERSIBLE REACTION

A reversible reaction (symbolised $\rightleftharpoons$) is a chemical change which, depending upon the conditions, may take place in either direction and reach a position of dynamic equilibrium in which the rate of the forward reaction equals the rate of the reverse action.
◀ Equilibrium ▶

R$_f$ VALUE

The R$_f$ value of a component in a mixture being separated by **chromatography** is the distance travelled by the component divided by the distance travelled by the solvent front. For a specified chromatographic system at a given constant temperature, each component has a characteristic R$_f$ value which may be used to identify the component.
◀ Paper chromatography, Thin-layer chromatography ▶

SALT BRIDGE

A salt bridge is an electrolytic connection between the solutions of two half-cells which minimises the liquid junction potential in the complete electrochemical cell. The single 'salt' in the bridge is usually saturated aqueous ammonium nitrate, potassium chloride or potassium nitrate because the cation and anion have similar transport numbers. In order to minimise diffusion of the salt into the half-cell electrolytes, the 'bridge' supporting the saturated aqueous salt may be an agar gel set in an inverted U-tube or simply a strip of filter paper.

◀ Electrochemical cell, Junction potential ▶

SATURATED HYDROCARBON

A saturated hydrocarbon is a compound, of carbon and hydrogen only, in which the atoms are joined only by single covalent bonds so that the molecule does not contain double or triple bonds. The two broad classes of acyclic and alicyclic hydrocarbons are the alkanes and cycloalkanes.

◀ Alkanes ▶

SATURATED SOLUTION

A saturated solution is a solution that is in dynamic equilibrium with undissolved solute. In a saturated aqueous solution of a sparingly soluble electrolyte X_xY_y at constant temperature, $[X^{y+}(aq)]^x[Y^{x-}(aq)]^y$ has a constant value known as the solubility product of the electrolyte.

s-BLOCK ELEMENTS

The s-block elements (Fig. S.1) are the Group 1 (alkali metals) and Group II (alkaline earth metals) elements on the left of the periodic table. The elements show many common properties because in the ground state their atoms have either one or two electrons in the outer s-orbital.

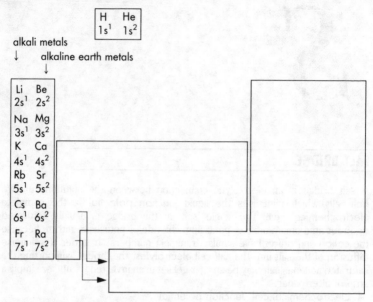

Fig. S.1 s-block elements and hydrogen

PHYSICAL PROPERTIES

The s-block elements are ductile, malleable, electrically and thermally conducting metals. They are softer than most transition metals and have lower densities, melting points, boiling points and standard enthalpy changes of melting and of boiling. Hardness, density, and enthalpies of melting and boiling all decrease with increasing atomic number in each group. The metallic bonding is weaker in the s-block elements than in most transition metals, and its strength decreases as the atomic number and size of the atom increases down each group.

CHEMICAL PROPERTIES

The s-block metals are powerful reducing agents. The atoms lose their outer s-subshell electrons to form very stable cations with the same electronic configurations as their corresponding noble gases – for instance:

$$Na(2,8,1) \rightarrow Na^+(2,8) + e^- \qquad Mg(2,8,2) \rightarrow Mg^{2+}(2,8) \qquad [Ne\ 2,8]$$

In their compounds, which are normally white or colourless, the s-block elements have fixed oxidation states of +1 for Group I and +2 for Group II. The elements burn to form oxides (such as Li_2O), peroxides (Na_2O_2) or superoxides (KO_2), depending upon their reactivity. The s-block metals combine directly with halogens, hydrogen and sulphur to form halides, hydrides and sulphides:

$$2Na(s) + Cl_2(g) \xrightarrow{\text{heat}} 2NaCl(s)$$

$$Ca(s) + H_2(g) \xrightarrow{\text{heat}} CaH_2(s)$$

$$Mg(s) + S(s) \xrightarrow{\text{heat}} MgS(s)$$

The metals react with water to form hydrogen and **hydroxides** (or oxides) that are strongly basic:

$$2Na(s) + 2H_2O(l) \longrightarrow 2Na^+(aq) + 2OH^-(aq) + H_2(g)$$

$$Mg(s) + H_2O(g) \xrightarrow{\text{heat}} MgO(s) + H_2(g)$$

The **aqueous** s-block metal cations are so stable that they do not discharge at the **cathode** during **electrolysis**. Consequently, these electropositive metals have to be extracted by electrolysis of their molten anhydrous salts.

SECONDARY ALCOHOL

A secondary alcohol is an alcohol containing the group >CHOH. Secondary alcohols differ from primary and tertiary alcohols by forming **ketones** when oxidised by **oxidising agents** such as acidified aqueous potassium dichromate(VI). **Primary alcohols** may be oxidised partially to aldehydes or completely to **carboxylic acids**, but tertiary alcohols are very resistant to oxidation.

◀ Alcohols ▶

SECOND-ORDER REACTION

A second-order reaction is a reaction with a differential **rate equation** of the form $-d[R]/dt = k[R]^2$, an integrated rate equation of the form $1/[R] = kt + 1/[R]_i$ and a half-life, $t_{1/2} = 1/k[R]$, dependent upon the concentration. $[R]_i$ and $[R]$ represent the **concentration** of a reactant R initially (at $t = 0$) and at time t, respectively; k is the **rate constant**; $-d[R]$ dt, the rate of decrease of the concentration of a reactant with time, represents the rate of reaction which, in a second-order reaction, is directly proportional to the concentration raised to the second power (see Fig. S.2).

The gas-phase decomposition of nitrogen dioxide into nitrogen monoxide and oxygen follows second-order kinetics, although decompositions of gases are often **first-order reactions**. The gas-phase combination of hydrogen and iodine to form hydrogen iodide is first-order with respect to hydrogen, first-order with respect to iodine vapour and second-order overall – that is:

rate = $k[H_2(g)][I_2(g)]$

◀ Rate of reaction ▶

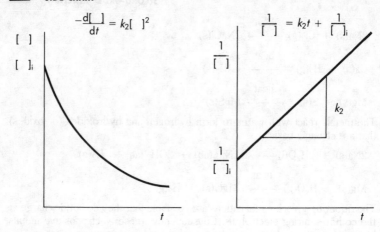

Fig. S.2 Second-order reaction kinetics

SIDE-CHAIN

A side-chain is a small alkyl group attached to a larger hydrocarbon chain or ring. A **hydrocarbon** with one or more side-chains attached to the main carbon chain is called a *branched-chain hydrocarbon*. The greater the number of side-chains, the greater is the degree of branching. Branched-chain **isomers** of a straight-chain hydrocarbon have lower boiling points because the side-chains keep the main chains further apart and hinder the operation of **van der Waals** attractive forces between the main chains.

SINGLE BOND

A single **covalent** bond is the net nuclear attraction for two **valence electrons** being shared between two **atoms**. A single bond can be represented in a diagram by a 'dot and cross' ($\dot{X}$) or a 'dash' ($-$) if each atom supplies one electron or by two 'dots' (:) or an arrow ($\leftarrow$) if one atom supplies both electrons to form a dative or coordinate bond. The most usual representation of a single bond is the 'dash' or single line drawn between the symbols of the two covalently bonded atoms: H–Cl, for instance.
◄ Covalent bonding ►

S_N1 REACTION

An S_N1 reaction is a **unimolecular** (1) **nucleophilic** ($_N$) **substitution** (S) reaction.

The **hydrolysis** of 2-bromo-2-methylpropane by aqueous sodium hydroxide is an S_N1 reaction. '1' means that the **rate-determining step** involves only *one* species (one molecule of 2-bromo-2-methylpropane undergoing **heterolytic fission**) – see Fig. S.3.

Bromoalkane RBr	Rate equation	Mechanism
CH_3 $\|$ CH_3— C — Br $\|$ CH_3 2-bromo- 2-methyl-propane	rate α [RBr]	S_N1 (a two-step mechanism) $RBr \xrightarrow{\text{slow}} R^+$ $R^+ + OH^- \xrightarrow{\text{fast}} ROH$

Fig. S.3 Hydrolysis of bromoalkanes: S_N1 reaction
◄ Reaction mechanism ►

S_N2 REACTION

An S_N2 reaction is a **bimolecular** (2) **nucleophilic** ($_N$) **substitution** (S) reaction.

The **hydrolysis** of 1-bromobutane by aqueous sodium hydroxide is an S_N2 reaction. '2' means that the **rate determining step** involves *two* species (one molecule of the electrophilic 1-bromobutane AND one nucleophilic **hydroxide ion** or water molecule) – see Fig. S.4.

Bromoalkane RBr	Rate equation	Mechanism
H $\|$ $CH_3CH_2CH_2$— C — Br $\|$ H 1-bromobutane	rate α [RBr][OH$^-$]	S_N2 (a one-step mechanism) $RBr + OH^- \xrightarrow{\text{slow}} ROH + Br^-$

Fig. S.4 Hydrolysis of bromoalkanes: S_N2 reaction
◄ Reaction mechanism ►

SOAPS

Soaps are anionic **detergents** consisting mainly of the sodium salts of **long-chain fatty acids**. They are made by alkaline **hydrolysis** of natural **fats** and **oils** in a process called *saponification* (Latin *sapo* – soap). The **triglycerides** are boiled with **aqueous** sodium hydroxide and, when hydrolysis is complete, the soap and glycerine are separated:

$$CH_3(CH_2)_{16}CO_2CH_2$$
$$CH_3(CH_2)_{16}CO_2CH \quad + \quad 3NaOH(aq) \rightarrow 3CH_3(CH_2)_{16}CO_2^-Na^+ \quad + \quad CH_2OH$$
$$CH_3(CH_2)_{16}CO_2CH_2 \qquad\qquad\qquad\qquad\qquad\qquad\qquad\qquad CHOH$$
$$CH_2OH$$

stearin   caustic soda   sodium stearate   glycerine

Unlike the sodium salts, the calcium and magnesium salts of long-chain fatty acids are insoluble in water. Consequently, a soap such as sodium stearate is partly wasted when used in hard water because it produces a 'scum' of insoluble calcium stearate. Soapless detergents are not affected in the same way.

SOLUBILITY

The solubility of a substance is the minimum amount or mass of the substance required to produce a **saturated solution** in a specified mass or volume of solvent at a given temperature. Solubilities can be expressed in various **concentration** units, including **moles** of solute per 100 g water at 298 K. In general, the solubility of solids in liquids increases with increasing temperature, but the solubility of gases in liquids decreases with increasing temperature.

SOLUBILITY PRODUCT

The solubility product (K_{sp}) of a sparingly soluble electrolyte (X_xY_y) is given by $K_{sp} = [X^{y+}(aq)]^x[Y^{x-}(aq)]^y$, where $[X^{y+}(aq)]$ and $[Y^{x-}(aq)]$ represent the concentrations of the **cation** and **anion** respectively.

Values for K_{sp} are usually obtained by determining the very low ion **concentrations** using e.m.f. measurements. The units of K_{sp} depend upon the values of x and y. For the silver halides, AgCl, AgBr and AgI, the values of K_{sp} at 298 K are 2×10^{-10}, 5×10^{-13} and 8×10^{-18} mol^2 dm^{-6} respectively.

For substances of similar formula, K_{sp}, has the same units, and the smaller the value of K_{sp}, the less soluble the substance.
◄ Equilibrium constant ►

SOLVATION

Solvation is a term which refers to the attachment of **polar solvent molecules** to the **ions** of a solute by **ion—dipole electrostatic forces of attraction** to form solvated ions.
◄ Hydration ►

SOLVENT EXTRACTION

Solvent extraction is the technique of separating one or more components from a mixture using a liquid that dissolves those components but leaves the rest of the mixture unaffected. When the solvent percolates through a mixture of solids, the extraction process is usually known as *leaching*, particularly if water is the solvent.

The solution mining of sodium chloride from underground salt deposits, by pumping water into them and brine out of them, could be regarded as solvent extraction on the grand industrial scale. And the making of a cup of coffee using freshly ground beans in a percolator is leaching on a small domestic scale. Solvent extraction is widely used in the preparation and purification of a range of natural oils and other products. When the mixture of components is in the liquid phase, the extracting solvent must be immiscible with it.

Ether extraction is an important laboratory technique that uses ethoxyethane to extract organic compounds of high **molar mass** and low volatility from an **aqueous** mixture of products.

◄ Distribution coefficient, Partition coefficient ►

SPECTATOR ION

Spectator ion is the name given to **anions** and **cations** which accompany a reaction but do not take part in it. When **aqueous** silver nitrate is added to aqueous sodium chloride, a white precipitate of silver chloride is immediately formed:

$$AgNO_3(aq) + NaCl(aq) \rightarrow AgCl(s) + NaNO_3(aq)$$

The nitrate anions and sodium cations remain in solution as aqueous ions. Consequently, these spectator ions are frequently ignored and the ordinary equation written as an ionic equation:

$$Ag^+(aq) + Cl^-(aq) \rightarrow AgCl(s)$$

Although spectator ions appear not to take part in a reaction, their nature and concentration may still influence the reaction. For example, the ionic equation for the reaction of any **strong acid** with any **strong base** may well be $H_3O^+(aq) + OH^-(aq) \rightarrow 2H_2O(l)$ but the **enthalpy change** is not quite the same in all cases.

◄ Ionic equation ►

SPECTROSCOPY

Spectroscopy is the collective term for instrumental techniques in analytical chemistry in which measurements of electromagnetic radiation, absorbed or emitted by energy changes in a substance, yield information about the composition and structure of the substance.

Absorption spectroscopy originated with the German glass-maker, Josef Fraunhofer in 1814 when he discovered dark lines in the solar spectrum. About twenty-five years later, the French physicist, Jean Bernard Leon Foucault, showed that these dark lines matched the bright lines in the spectrum from a sodium flame. *Emission spectroscopy* was launched by the work of the German scientists, R. Bunsen and G.R. Kirchoff, who established the law connecting the emission and absorption of light, and who discovered rubidium and caesium while studying the emission spectra of lithium, sodium and potassium.

◄ Infra-red absorption spectroscopy, Nuclear magnetic resonance, Ultraviolet spectroscopy ►

STABILITY CONSTANT

Stability constants are **equilibrium constants** for reversible **ligand** exchange reactions involving metal ion **complexes**.

The law of chemical equilibrium can be applied to the reaction which takes place when ammonia molecules (strong **Lewis base**) replace water molecules (weaker Lewis base) from tetraaquacopper(II) complex cations to form the more stable tetraamminecopper(II) complex cations:

$$4NH_3(aq) + Cu^{2+}(aq) \rightleftharpoons Cu(NH_3)_4^{2+}(aq)$$

$$\frac{[Cu(NH_3)_4^{2+}(aq)]}{[NH_3(aq)]^4[Cu^{2+}(aq)]} = K = 1.3 \times 10^{13} \text{ mol}^{-4} \text{ dm}^{12}$$

K is the stability constant of the tetraamminecopper(II) complex cation. Stability constants are usually large and therefore handled as the logarithms of their values (see Fig. S.5).

Ligand	Aqueous complex ion	Stability constant	Log K
Cl^-	$CuCl_4^{2-}$	4.2×10^5	5.6
NH_3	$Cu(NH_3)_4^{2+}$	1.3×10^{13}	13.1
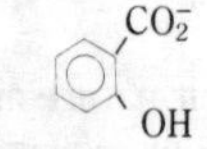 (benzene ring with CO_2^- and OH)	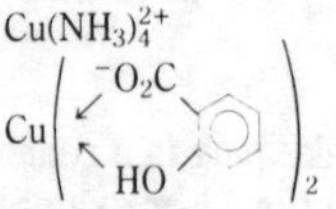 $Cu\left(\substack{^-O_2C \\ HO}\right)_2$	7.9×10^{16}	16.9
$EDTA^{4-}$	$Cu(EDTA)^{2-}$	6.3×10^{18}	18.8
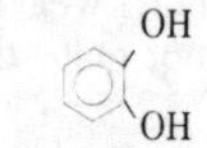 (benzene ring with OH and OH)	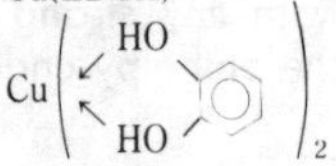 $Cu\left(\substack{HO \\ HO}\right)_2$	1.0×10^{25}	25.0
NH_3	$Co(NH_3)_6^{2+}$	2.5×10^4	4.4
NH_3	$Ni(NH_3)_6^{2+}$	1.0×10^8	8.0
NH_3	$Ag(NH_3)_2^+$	1.7×10^7	7.2

Fig. S.5 Stability constants of complex ions

Complexes formed by polydentate ligands have higher stability constants than complexes formed by monodentate ligands. From a comparison of the stability constants, it can be predicted that **EDTA** (a hexadentate ligand) will displace ammonia molecules (monodentate ligands) from $Cu(NH_3)_4^{2+}(aq)$:

$$ETDA^{4-}(aq) + Cu(NH_3)_4^{2+}(aq) \rightarrow Cu(EDTA)^{2-}(aq) + 4NH_3(aq)$$

and that

$$\frac{[Cu(EDTA)^{2-}(aq)][NH_3(aq)]^4}{[EDTA^{4-}(aq)][Cu(NH_3)_4^{2+}(aq)]} = \frac{\text{stability constant } Cu(EDTA)^{2-}(aq)}{\text{stability constant } Cu(NH_3)_4^{2+}(aq)}$$

◀ Complex, Equilibrium constant ▶

STANDARD ELECTRODE

A standard electrode is an **electrochemical half-cell** taken as a reference under standard conditions of **concentration** 1 mol kg^{-1} (1 mol dm^{-3}), pressure 101.325 kPa (1 atm) and temperature 298.15 K (25°C). The values shown in parentheses are usually sufficient for A-level chemistry.

The **standard hydrogen electrode** is the primary reference, so that its electrode potential is taken as zero. However, the hydrogen electrode is not very convenient for everyday practical work, so other electrodes such as the standard calomel electrode, are used as a secondary reference.

◀ Electrochemical cell, Electrode, Hydrogen electrode ▶

STANDARD ELECTRODE POTENTIAL

◀ Electrode potential ▶

STANDARD e.m.f.

The standard e.m.f. (electromotive force) of an **electrochemical cell** is the maximum potential difference (voltage) between the **electrodes** under standard conditions of **concentration** 1 mol kg^{-1} (1 mol dm^{-3}), pressure 101.325 kPa (1 atm) and temperature 298.15 K (25°C).

◀ Electromotive force ▶

STANDARD ENTHALPY CHANGE

◀ Enthalpy change ▶

STANDARD SOLUTION

A standard solution is a solution of accurately known **concentration** used in analytical chemistry for titrimetric and **volumetric analysis**.

◀ Titration, Volumetric analysis ▶

STANDARD STATE

The standard state of an element or compound is defined as the most stable physical form at a pressure of 101.325 kPa (1 atm) and a temperature of 298.15 K (25°C). The values shown in brackets are usually sufficient for A-level chemistry.

◀ Standard enthalpy change, Standard free energy change ▶

STEAM DISTILLATION

Steam distillation (Fig. S.6) is a technique for separating a mixture of water and organic compounds from involatile substances and for distilling organic compounds that have high molar masses, low volatilities and a tendency to decompose at their normal boiling points.

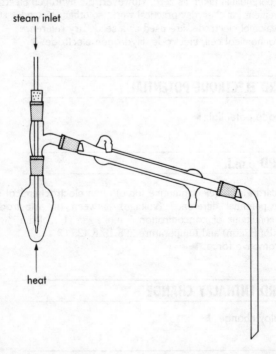

steam inlet

heat

Fig. S.6

Steam distillation works for organic compounds that are immiscible with water. Such compounds and water do not form a solution. They remain as two separate liquid phases, each of which separately contributes its full (pure) **vapour pressure** ($p_{organic}$ and p_{water}) at a given temperature to the total vapour pressure (p) of the mixture at that given temperature; i.e.

$p_{\text{mixture}} = p_{\text{organic}} + p_{\text{water}}$. Consequently, the liquid mixture will boil at a temperature lower than that of either the water or the organic. For example, at an atmospheric pressure of 760 mm Hg, a mixture of an immiscible organic compound and water will distil at 97°C if the compound has a vapour pressure of 78 mm Hg at 97°C, because the vapour pressure of water is 682 mm Hg at 97°C. The relative amounts of water and organic in the steam distillate can be estimated by treating the vapours as perfect gases, applying the ideal gas equation and using **Dalton's law of partial pressures** to arrive at the following expression:

$$\frac{\text{mass water}/18}{\text{mass compound}/M} = \frac{p_{\text{water}}}{p_{\text{organic}}}$$

where 18 and M are the relative molecular mass of water and the organic compound respectively, p_{water} and p_{organic} are the partial pressures of water and the organic compound respectively, in the vapour at equilibrium with the liquid mixture.

STEREOCHEMICAL FORMULA

A stereochemical formula (see Fig. S.7) shows the three-dimensional spatial arrangement of bonds, atoms and groups in a molecule.
◀ Geometric isomerism, Nuclear isomerism, Optical isomerism ▶

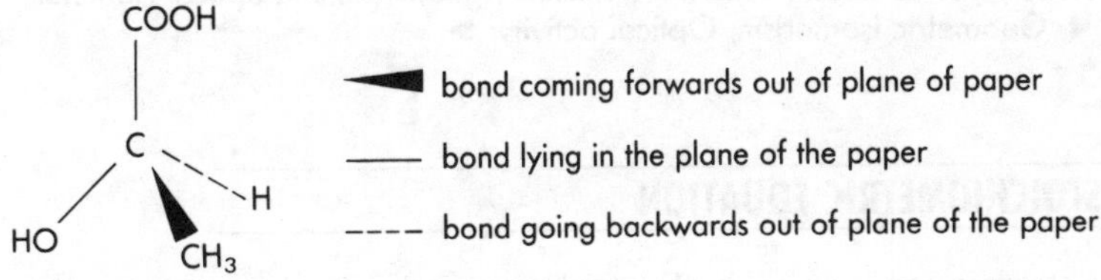

bond coming forwards out of plane of paper

bond lying in the plane of the paper

bond going backwards out of plane of the paper

Fig. S.7 Convention for a stereochemical formula

STEREOISOMERISM

nuclear isomers

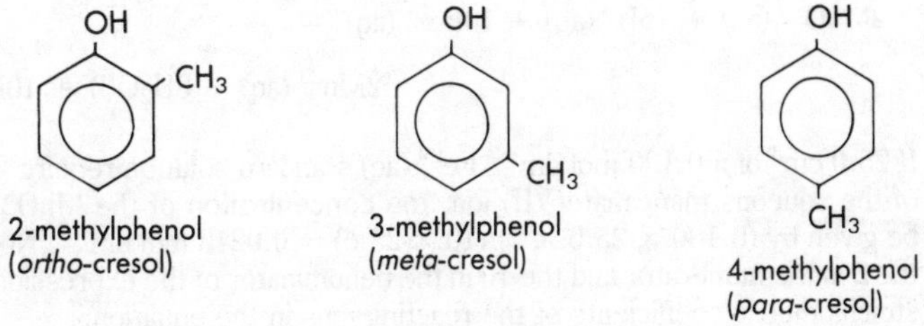

2-methylphenol
(*ortho*-cresol)

3-methylphenol
(*meta*-cresol)

4-methylphenol
(*para*-cresol)

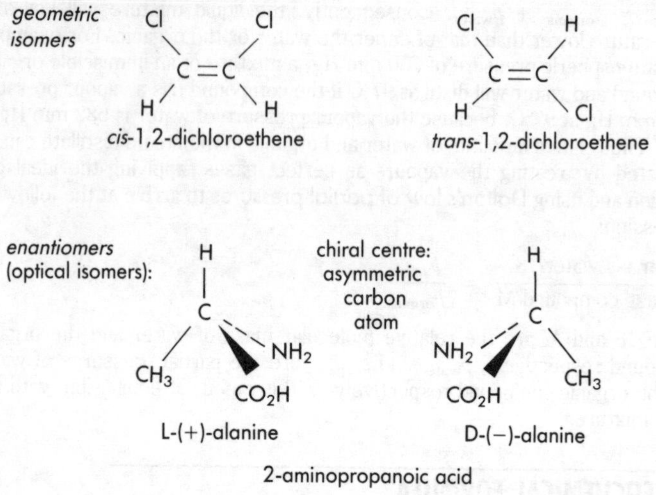

geometric isomers

cis-1,2-dichloroethene

trans-1,2-dichloroethene

enantiomers (optical isomers):

chiral centre: asymmetric carbon atom

L-(+)-alanine

D-(−)-alanine

2-aminopropanoic acid

Fig. S.8 Types of stereoisomerism

Stereoisomerism refers to the existence of **isomers** having in their molecular structure the same **groups** of **atoms** differently arranged in space. The three broad types of stereoisomers are **nuclear**, **geometric** and **optical** isomers.
◀ Geometric isomerism, Optical activity ▶

STOICHIOMETRIC EQUATION

A stoichiometric equation is a balanced equation showing the relative amounts of reactants and products. It usually has simple whole numbers which precede the formulae. These are known as *stoichiometric coefficients*.

Calculations based on chemical reactions depend upon the stoichiometric equation for the reaction. For example, in order to standardise aqueous potassium manganate(VII), the solution could be used to titrate a standard solution of ammonium iron(II) sulphate in aqueous sulphuric acid. The stoichiometric ionic equation for the **titration** is:

$$2MnO^-_4(aq) + 16H^+(aq) + 10Fe^{2+}(aq) \rightarrow$$

$$2Mn^{2+}(aq) + 8H_2O(l) + 10Fe^{3+}(aq)$$

If 25.0 cm³ of a 0.100 mol dm⁻³ Fe^{2+}(aq) standard solution require 23.5 cm³ of the aqueous manganate(VII) ion, the **concentration** of the MnO^-_4(aq) will be given by $(0.100 \times 25.0 \times 2)/(10 \times 23.5) = 0.0213$ mol dm⁻³. Notice that the 2 in the numerator and the 10 in the denominator of the expression are the stoichiometric coefficients of the reacting ions in the equation.

STP

STP (or s.t.p.) is the abbreviation for standard temperature and pressure as applied to gases and refers to 273.15 K (0°C) and 101.325 kPa (1 atm). At STP the **molar volume** of an ideal gas is taken to be 22.413 dm³. For A-level chemistry calculations, STP is usually taken as 273 K and 101 kPa and the molar gas volume at STP is taken as 22.4 dm³.

◀ Molar volume ▶

STRAIGHT CHAIN

A straight chain is an open chain of **atoms** with no atom bonded to more than two other atoms.

Pentane, $CH_3CH_2CH_2CH_2CH_3$, is the fifth member of the **homologous series** of straight chain alkanes, C_nH_{2n+2}. The carbon atoms at the end of the chain are bonded to one other carbon atom. The carbon atoms in the chain are bonded to two other atoms. Although the displayed formula may seem to suggest that the chain is 'straight', it must be remembered that the bonds to each carbon atom are directed towards the corners of a tetrahedron and that rotation can occur about the **single bonds**.

STRONG ACID

A strong **acid** is an acid that is completely ionised in water.

◀ Acid dissociation constant ▶

STRONG BASE

A strong **base** is a base that is completely ionised in water.

◀ Base dissociation constant ▶

STRUCTURAL FORMULA

A structural **formula** is a formula that shows the sequence and arrangement of **atoms** and **groups** in a **molecule**. It can often be written in more than one way. For example, hexanoic acid is a **fatty acid** with a **straight-chain** of six carbon atoms. $CH_3CH_2CH_2CH_2CH_2CO_2H$ and $CH_3(CH_2)_4CO_2H$ are acceptable formulae showing the structure of the chain. −COOH is a common alternative for the structural formula of the carboxyl group. $C_5H_{11}CO_2H$ is not satisfactory as a structural formula, because the C_5H_{11} does not distinguish between the hexyl, 2-methylpentyl or 3-methylpentyl groups. A structural formula should be unambiguous. The conventions should be carefully followed to avoid confusion, for example, between **alcohol** and **aldehyde functional groups** and between benzene and cyclohexane rings – see Fig. S.9.

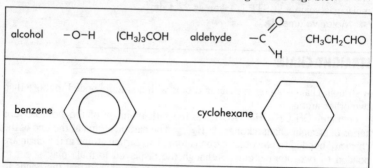

Fig. S.9 Structural formulae frequently confused

◀ Formula ▶

SUBSTITUENT

Substituent is the name given to an **atom** or **group** which has replaced another atom or group in a **molecule**. In a substituted **aromatic** compound, the substituent has an important influence on the response of the benzene ring to further attack by an **electrophile**. On the one hand, the compound forms mainly *1,2-* and *1,4-* subtitution products and reacts faster than benzene if the substituent already attached to the ring is electron-donating: $-CH_3$, $-OH$, $-NH_2$, for example. On the other hand, the compound forms mainly *1,3-*substitution products and reacts slower than benzene if the group already attached to the ring is electron-attracting: $-NO_2$, $-CO_2H$, $-SO_3H$, for example.

There are only **single bonds** in the electron-donating substituents but there is at least one **double bond** in the electron-attracting substituents. Whether a substituent is electron-donating or -attracting depends upon the **electronegativities** of its atoms and upon the interaction of their valence electrons with the delocalised π-electrons of the benzene ring. When the substituent is just a **halogen** atom, the balance between these two factors leads to the halogenobenzene compound forming mainly *1,2-* and *1,4-*substitution products, but reacting more slowly than benzene.

The methyl group in methylbenzene is an electron-donating substituent, so the benzene ring is readily attacked by electrophiles such as the nitronium ion, NO_2^+, to give various products, including three isomeric nitromethylbenzene compounds (see Fig. S.10).

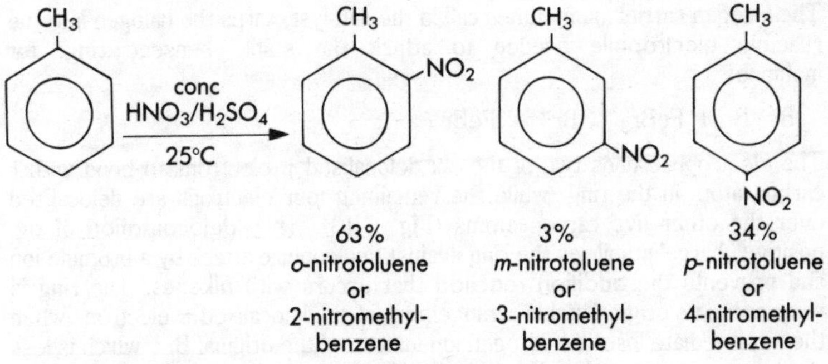

Fig. S.10 Isomeric substitution products of methylbenzene

◀ Electrophilic substitution ▶

SUBSTITUTION

A substitution reaction is a reaction in which one **atom** (or group) in a molecule is replaced by another atom (or group). In a **nucleophilic substitution**, the arriving (and departing) atom (or group) is an electron-pair donor – see Fig. S.11. In an **electrophilic** substitution, the attacking (and leaving) atom (or group) is an electron-pair acceptor – see Fig. S.12.

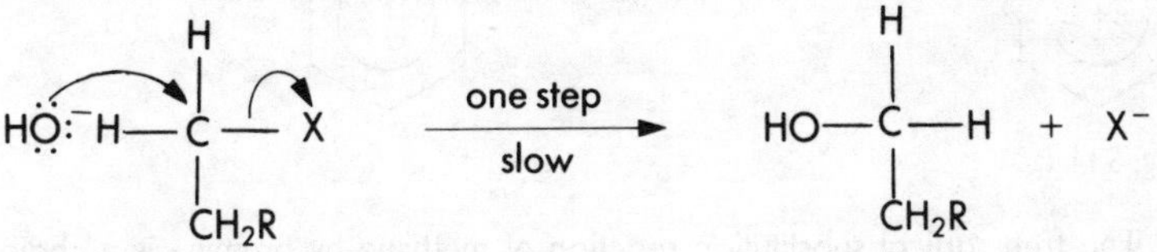

Fig. S.11 Nucleophilic substitution

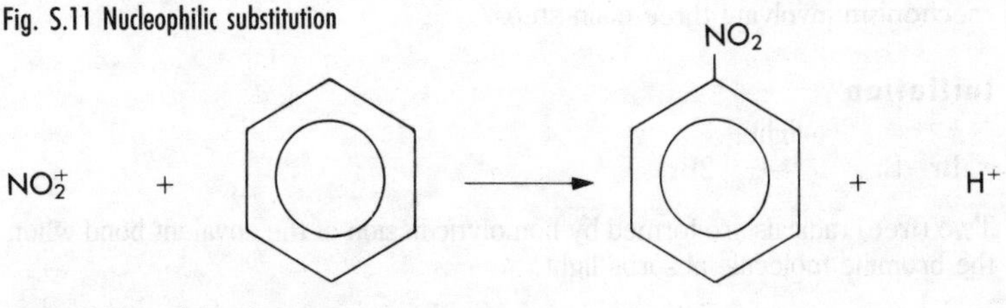

Fig. S.12 Electrophilic substitution

The reactions of benzene and its derivatives are characterised by electrophilic substitution reactions. Benzene itself reacts with bromine or chlorine in the presence of a **halogen carrier** to form bromo- or chloro-benzene:

$$C_6H_6 + Br_2 \xrightarrow[\text{Fe(s) or FeBr}_3]{Br_2(l)} C_6H_5Br + HBr$$

The halogen carrier, sometimes called the catalyst, turns the halogen into the reactive **electrophile** needed to attack the stable benzene ring; for instance:

$$Br–Br + FeBr_3 \rightarrow Br^+ + FeBr_4^-$$

The electrophile uses two of the six delocalised π-electrons to bond with a carbon atom in the ring, while the remaining four electrons are delocalised over the other five carbon atoms (Fig. S.13). This 'delocalisation of the positive charge' stabilises the ring against nucleophilic attack by a bromide ion and prevents the **addition reaction** that occurs with **alkenes**. The ring is restored to its original stable arrangement of six delocalised π-electrons when the intermediate loses a hydrogen ion instead of the original Br^+ which is less stable/more reactive than H^+ (Fig. S.14).

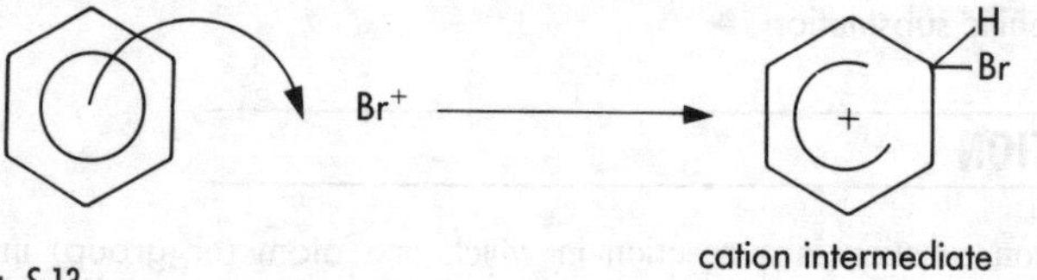

Fig. S.13

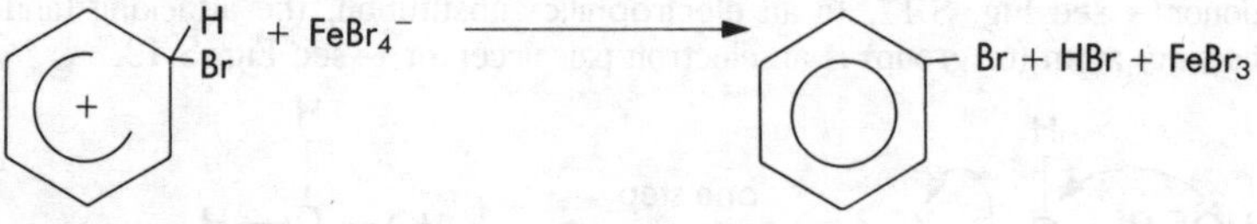

cation intermediate

Fig. S.14

The **free-radical substitution reaction** of methane by bromine is a **chain mechanism** involving three main steps.

Initiation

$$Br–Br \xrightarrow{\text{light}} 2Br\cdot$$

Two (free) radicals are formed by homolytic fission of the covalent bond when the bromine molecule absorbs light.

Propagation

$$Br\cdot + H–CH_3 \rightarrow H–Br + \cdot CH_3$$

The bromine radical attacks the methane molecule to produce a methyl radical and a hydrogen bromide molecule.

$$Br–Br + \cdot CH_3 \rightarrow Br–CH_3 + \cdot Br$$

The methyl radical attacks a bromine molecule to produce a bromine radical and a bromomethane molecule.

Termination

$$\cdot CH_3 + \cdot CH_3 \rightarrow CH_3-CH_3$$

Two methyl radicals form an ethane molecule.

The net effect of the two propagation steps is: $Br_2 + CH_4 \rightarrow HBr + CH_3Br$. However, a free-radical substitution reaction produces several compounds, because a substitution product may itself participate in the propagation step:

$$\begin{array}{cccc}
\overset{\displaystyle Br}{\underset{\displaystyle H}{H-\overset{|}{\underset{|}{C}}-Br}} &
\overset{\displaystyle Br}{\underset{\displaystyle H}{H-\overset{|}{\underset{|}{C}}-Br}} &
\overset{\displaystyle Br}{\underset{\displaystyle Br}{H-\overset{|}{\underset{|}{C}}-Br}} &
\overset{\displaystyle Br}{\underset{\displaystyle Br}{Br-\overset{|}{\underset{|}{C}}-Br}}
\end{array}$$

Homolytic free-radical chlorination of methane (CH_4) and of the methyl group in methylbenzene ($C_6H_5-CH_3$ toluene) gives CH_3Cl, CH_2Cl_2, $CHCl_3$, CCl_4 and $C_6H_5-CH_2Cl$, $C_6H_5-CHCl_2$, $C_6H_5-CCl_3$ respectively.
◀ S_N1 reaction, S_N2 reaction ▶

SUBSTRATE

Substrate is the name for any substance which an **enzyme** acts upon. Urease is an enzyme which catalyses the hydrolysis of **urea**. Urea is the substrate of the enzyme urease.

SULPHATES

A sulphate is an inorganic salt of sulphuric acid containing the tetrahedral ion SO_4^{2-}.

The solubility of the sulphates in water varies from the very soluble **alkali metal** sulphates to the highly insoluble barium sulphate. Aqueous barium chloride or nitrate is used to test for sulphate ions in aqueous solution by the formation of a white precipitate that is insoluble in hydrochloric or aqueous nitric acid. **Esters** formed by sulphonating alcohols are called *organic sulphates*. Sodium (lauryl) dodecylsulphate $CH_3(CH_2)_{10}CH_2O-SO_3^-Na^+$ is an important anionic soapless **detergent**. Metal sulphates can be made by reacting an excess of the metal **oxide**, **hydroxide** or **carbonate** with sulphuric acid. If an excess of the acid is used, the hydrogensulphate is formed:

$$2NaOH(aq) + H_2SO_4(aq) \rightarrow Na_2SO_4(aq) + 2H_2O(l)$$

$$NaOH(aq) + H_2SO_4(aq) \rightarrow NaHSO_4(aq) + H_2O(l)$$

SULPHIDES

A sulphide is a compound of an element with sulphur. Metal sulphides are regarded as the salts of H_2S. Most metal sulphides are insoluble in water and are precipitated when hydrogen sulphide is bubbled into an aqueous metal salt solution. At one time a system of **qualitative analysis** incorporated the formation and appearance of sulphide precipitates as part of the techniques for detecting and identifying metallic elements in compounds and mixtures. This system has largely been replaced by more rapid and sensitive instrumental methods.

SULPHITES

A sulphite is an inorganic salt of sulphurous acid, H_2SO_3. Sulphites are formed by the reaction of sulphur dioxide with alkalis. They are reducing agents used, for example, as antioxidants in food preservation. The sulphite ion is readily oxidised to the sulphate ion quantitatively. Consequently, a standard solution of aqueous manganate(VII) ions can be used in excess acid for the **quantitative analysis** of sulphites:

$$2MnO_4^-(aq) + 6H^+(aq) + 5SO_3^{2-}(aq) \rightarrow 2Mn^{2+}(aq) + 3H_2O(l) + 5SO_4^{2-}(aq)$$

The aqueous sulphite ion gives a white precipitate with aqueous barium ions but, unlike the white precipitate given by the aqueous sulphate ion, barium sulphite reacts with HCl(aq) or HNO_3(aq).

SULPHONATION

Sulphonation is an **electrophilic substitution reaction** which introduces a sulphonic group $-SO_3H$ into a compound.

Concentrated or 'fuming' sulphuric acid, chlorosulphonic acid and sulphur trioxide have all been used as sulphonating agents. The benzene ring in straight-chain alkylbenzene hydrocarbons can be sulphonated with a gaseous mixture of sulphur trioxide and air to produce alkylbenzene sulphonic acids on the industrial scale. The sodium salts of these acids are important biodegradable soapless **detergents**.

◀ Electrophilic substitution ▶

SUPERSATURATED SOLUTION

A supersaturated solution is a solution containing a solute with a higher **concentration** than it would have in a **saturated solution** at the same temperature. Supersaturated solutions may form when hot, dust-free saturated solutions are cooled. They are unstable and crystallise even when just one 'seed' crystal of the solute is added.

When NaOH(aq) or Na_2CO_3(aq) is added to an aqueous alkaline earth metal salt, supersaturated solutions are often obtained, because precipitates of the Group II **carbonates** and **hydroxides** are slow to form. In the case of such sparingly soluble **electrolytes** as calcium carbonate, a supersaturated solution would be described by the condition that the product of the concentration of the ions $[CaO^{2+}(aq)][CO_3O^{2-}(aq)]$ would have a higher value than the **solubility product** K_{sp}.

SYNDIOTACTIC POLYMER

A syndiotactic polymer is a polymer with a stereoregularity greater than that of atactic polymers, but less than that of isotactic polymers, with the groups (X) alternating on either side of the carbon chain:

◀ Isotactic polymer ▶

SYNTHESIS

Synthesis is the preparation of a compound from simpler compounds or from its elements. Synthetic organic chemistry began in 1824 with Wohler's synthesis of urea from inorganic chemicals. Synthetic polymer chemistry developed from the synthesis of phenol-formaldehyde resins in 1909, artificial rubber in 1914, polyethylene in 1933 and nylon in 1934.

In A-level practical chemistry, a variety of organic compounds can be prepared in the laboratory by a one- or two- stage synthesis. For example, an ester may be prepared from an alcohol by oxidising some of the alcohol to a carboxylic acid and then reacting together the carboxylic acid and the rest of the alcohol to form the ester. The industrial production of ammonia by the Haber process is the synthesis of an inorganic compound from its elements.

TAUTOMERISM

Tautomerism is a special case of **structural isomerism** in which the tautomers (tautomeric isomers) are directly interconvertible and capable of co-existing in **dynamic equilibrium** with each other. Carbonyl compounds may exist as pairs of tautomers, known as the *keto* and *enol* forms:

$$-CH_2-\underset{\underset{O}{\|}}{C}- \quad \rightleftharpoons \quad -CH=\underset{\underset{OH}{|}}{C}-$$

keto-form   *enol*-form

The simple **aldehydes** and **ketones** such as ethanal and propanone exist almost entirely in the *keto*-form, CH_3CHO and $(CH_3)_2CO$, rather than the *enol*-form $CH_2=CH(OH)$ and $CH_2=C(OH)CH_3$.

◀ Isomerism ▶

TERMINATION

The termination stage in a **radical chain reaction** is the third stage, in which free **radicals** required for the **propagation** stage take part in reactions that do not regenerate the free radicals.

A termination step usually involves the combination of two free radicals. For example, in the free-radical bromination of methane, three possible termination steps would be the combination of two methyl radicals to form an ethane molecule ($\cdot CH_3 + \cdot CH_3 \rightarrow CH_3-CH_3$), a bromine radical and methyl radical to form a bromomethane molecule ($Br\cdot + \cdot CH_3 \rightarrow CH_3Br$) and two bromine radicals to reform a bromine molecule ($2Br\cdot \rightarrow Br_2$). The last example is simply the reverse of the **initiation** stage.

◀ Radical chain mechanism ▶

TERTIARY ALCOHOL

A teritary alcohol is an alcohol containing the group $\geqslant C-OH$. Tertiary alcohols differ from primary and secondary alcohols in being very resistant to **oxidation** by **oxidising agents** such as acidified aqueous potassium dichromate(VI). Primary alcohols may be oxidised partially to aldehydes or completely to carboxylic acids. Secondary alcohols are oxidised to ketones.

◄ Alcohol ►

TERYLENE

Terylene is a trade name for polyesters derived from 1,4-benzene-dicarboxylic (terephthalic) acid and propan-1,2-diol (glycol).

◄ Polyester ►

THERMOCHEMISTRY

Thermochemistry is the branch of physical chemistry concerned with the study of heat changes which accompany chemical and physical processes.

◄ Bomb calorimeter, Enthalpy change, Flame calorimeter, Hess's law ►

THERMOPLASTIC POLYMER

A thermoplastic polymer is one which softens and can be moulded when heated and sets into the moulded shape when cooled. The nylons and the poly(alkene) addition polymers are two important classes of thermoplastic polymer.

THERMOSETTING POLYMER

A thermosetting polymer is one which polymerises and sets permanently into a rigid solid by cross-linking in a three-dimensional network. Phenol-formaldehyde and polyester resins are two important classes of thermosetting polymer.

THIN-LAYER CHROMATOGRAPHY

Thin-layer chromatography (TLC) is a technique to separate a mixture of solutes using liquids as the mobile phase and a very thin layer of powdered adsorbent, supported on a glass or rigid plastic sheet, as the stationary phase. The chromatographic plate, with its 0.1–0.2 mm thick adsorbent layer of alumina or silica gel, is prepared and handled in a similar way to the paper sheet in one-way and two-way paper chromatography. TLC is more widely used than paper chromatography because it is faster, more sensitive and more convenient. The composition of the adsorbent layer can be varied to suit particular mixtures, and a *fluorescer* can be incorporated in the layer so that the position of the spots can be seen in ultraviolet light.

◄ Paper chromatography ►

TITRATION

Titration is a technique in **volumetric analysis** for determining the **concentration** of solute by reacting the solution with a **standard solution** and measuring the volumes needed to complete the reaction.

The volume of *titrand* (solution to be analysed) is usually measured into a conical flask with a pipette. The volume of *titrant* (standard solution of known concentration) is usually run from a burette, and the volume delivered is measured by the difference in readings on the burette – see Fig. T.1. The **end-point** for the reaction is usually found by a change in the colour of an **indicator** which is present in the conical flask and is affected by the changing conditions in the solution. Titrations are often automated, using a change in, say, the electrical properties of the solution in the flask to reveal the end-point.

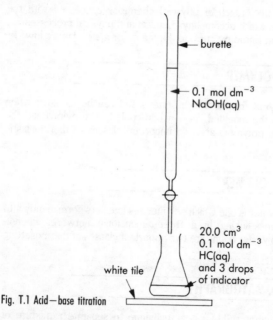

burette

0.1 mol dm^{-3}
NaOH(aq)

20.0 cm^3
0.1 mol dm^{-3}
HC(aq)
and 3 drops
of indicator

white tile

Fig. T.1 Acid–base titration

TITRATION CURVE

A titration curve is a graph of some property of a titrand plotted against the total volume of titrant added during a titration. For acid–base titrations, the titration curve is usually a graph of **pH** as the ordinate (y-axis) against total volume of **aqueous acid** or **alkali** added (see Fig. T.2). The shape of a pH-titration curve depends upon the strength of the acid and alkali and upon the **concentrations** of their solutions. A **dilute** solution of a **weak alkali** cannot be titrated with a dilute solution of a **weak acid** because the change in pH at

titration of 25.0 cm³ 0.1 mol dm⁻³ alkali in a conical flask by 0.1 mol dm⁻¹ HCl(aq) added from a burette

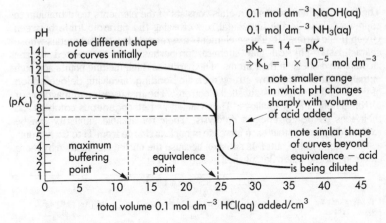

0.1 mol dm⁻³ NaOH(aq)
0.1 mol dm⁻³ NH₃(aq)
$pK_b = 14 - pK_a$
$\Rightarrow K_b = 1 \times 10^{-5}$ mol dm⁻³

note smaller range in which pH changes sharply with volume of acid added

note similar shape of curves beyond equivalence — acid is being diluted

Fig. T.2 pH-titration curve of strong and weak alkali by strong acid

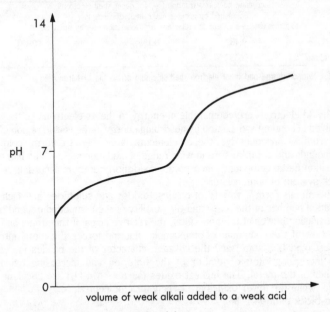

Fig. T.3 pH-volume curve for weak acid and weak base

the **equivalence point** is too small and too gradual to produce a sharp, detectable **end-point** (see Fig. T.3).

◀ Acid—base titration, Titration ▶

TRANSITION METAL

The first transition series of metals consists of the elements from titanium to copper inclusive in the horizontal row crossing the **periodic table** between Group II and Group III. These elements are regarded as typical metals whose characteristic physical and chemical properties may be related to the incomplete d-shells in their atoms. The hardness, high melting points and high enthalpies of melting show strong **metallic bonding**, involving **delocalisation** of d-subshell as well as s-subshell electrons. The metals are paramagnetic or ferromagnetic and form alloys. They can act as **heterogeneous catalysts** and their ions as **homogeneous catalysts**. Their remarkable similarity can be explained by saying that each increase in nuclear charge from Ti to Cu has only a small effect on the outer 4s subshell because the nucleus is shielded by more electrons in the inner 3d subshell – see Fig. T.4.

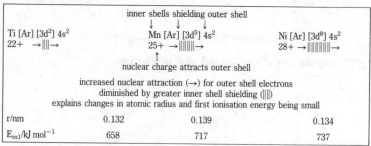

Fig. T.4 Nuclear charge and inner electron shell shielding across the first transition series

The 3d electrons are comparable in energy to the 4s electrons so they are involved in bonding and excited by absorbing energy in the visible region of the spectrum. Consequently, these elements are able to form coloured compounds and **complex ions** in wide range of oxidation states. The names of transition metal compounds incorporate **oxidation numbers** according to the Stock system of nomenclature.

The metals form a variety of halides, **oxides** and **sulphides** in which the oxidation number of the metal and the stability of the compound depend upon the oxidising power of the non-metal, the relative sizes of the atoms and the sturcture of the compound. For example, fluorine forces the metal into its highest oxidation state, and the covalent character of the halides increases with increasing **atomic number** of the **halogen** and increasing oxidation number of the metal. The highest oxides (such as Mn_2O_7) are covalent and acidic, while the lower oxides are largely ionic or neutral.

◄ d-block elements ►

TRANSITION METAL ION

Transition metals form **cations**, oxoanions and **complex** ions and display a

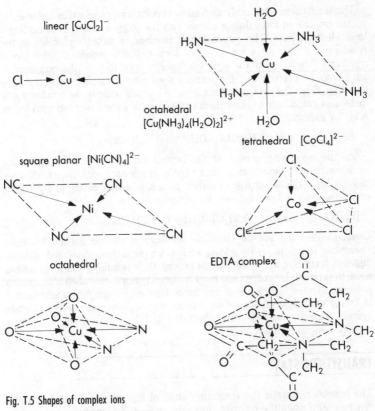

linear [CuCl₂]⁻

octahedral [Cu(NH₃)₄(H₂O)₂]²⁺

tetrahedral [CoCl₄]²⁻

square planar [Ni(CN)₄]²⁻

octahedral

EDTA complex

Fig. T.5 Shapes of complex ions

range of oxidation states. For A-level chemistry, the emphasis is upon the **aqueous** ions and the variety and relative stability of their oxidation states, the formation and relative stability of various complexes and the ability to act as **homogeneous catalysts** as their most important characteristics. The relative stability of the oxidation states is usually discussed in terms of standard electrode potentials, E^θ.

In the first transition series from Sc to Zn, the highest **oxidation number** reaches a maximum of seven with manganese and is related to the number of 3d and 4s electrons in the atom. The decrease in the range of oxidation states from manganese to zinc is related to the decrease in the number of unpaired 3d electrons in the atom. From left to right across the series Sc–Zn, the stability of the higher oxidation states decreases and that of the +2 oxidation state relative to the +3 state increases with atomic number. Compounds with elements in the intermediate oxidation states tend to **disproportionate**. The greater stability of Mn^{2+} (w.r.t. Mn^{3+}) and Fe^{3+} (w.r.t. Fe^{2+}) is related to the stability of the half-filled 3d-subshell. Transition metal ions tend to exist as **hydrated** cations in low-oxidation states and as oxoanions at high-oxidation states.

The small transition metal cation attracts polar water molecules and uses its vacant orbitals to form **dative bonds** with the oxygen atoms by accepting a **lone electron pair**. The **coordination number** of water molecules in the resulting complex aqua ion may be 2 or 4 (fairly common) or 6 (most common). The shape of the aqua ion depends upon the coordination number but cannot be predicted by the **valence shell electron pair repulsion theory** (see Fig. T.5). The central transition metal ion may polarise the bonded water molecules sufficiently to allow them to lose protons and make the aqua ion an acid; for example:

$$[Fe(H_2O)_6]^{3+}(aq) \rightleftharpoons [Fe(H_2O)_5OH]^{2+}(aq) + H^+(aq)$$

Since the **polarising power** of the central ion increases with increasing oxidation number, aqua cations with a triple charge are much more acidic than those with a double charge. Further proton loss can result in **hydrolysis** producing coloured precipitates; for instance:

$$[Fe(H_2O)_6]^{3+}(aq) \rightleftharpoons [Fe(H_2O)_3(OH)_3](s) + 3H^+(aq)$$

Consequently, the law of **chemical equilibrium** can be applied to these **reversible reactions**, i) by adding acid to suppress hydrolysis and stabilise aqueous transition metal salts when making their solutions and ii) by adding alkali to promote hydrolysis and form coloured precipitates when testing their solutions. Other complex ions are formed by exchanging the monodentate H_2O and OH^- **ligands** by stronger monodentate, bidentate or hexadentate ligands. Ligand exchange may affect the stability and colour of aqueous transition metal complex ions.

TRANSITION STATE

The transition state is the temporary state of highest energy achieved in a reaction step when the reactants form the **activated complex**.
◄ Activated complex, Activation energy ►

TRANSITION TEMPERATURE

The transition temperature is the temperature at which two different solid forms of the same substance can coexist in **equilibrium**. The rhombic and monoclinic enantiotropes of sulphur have a transition temperature of about 96°C. Below this temperature only the rhombic allotrope is stable. Above 96°C only the monoclinic form is stable.
◄ Allotropy, Allotrope ►

TRIGLYCERIDE

A triglyceride is an **ester** derived from **long-chain aliphatic carboxylic acids** and propane-1,2,3-triol (glycerine).
◄ Fat, Oil ►

TRIPLE BOND

A triple bond is a **covalent bond** formed between two atoms by the sharing of three pairs of **electrons**. The **isoelectronic diatomic molecules** of nitrogen (:N≡N:) and carbon monoxide (:C≡O:) contain triple bonds which are similar in length (0.110 nm and 0.113 nm) and strength (945 kJ mol^{-1} and 1077 kJ mol^{-1}) and which are shorter and stronger than their corresponding **double bonds**.

TROUTON'S RULE

The molar enthalpy change of vaporisation ($\triangle H^{\theta}_{evap}$ *or* $\triangle H^{\theta}_{b}$) of a liquid (at its normal boiling point) divided by its normal boiling point (in Kelvin) is constant ($\simeq 88$ J K^{-1} mol^{-1}).

This rule was put forward by Frederick Trouton in 1884. If the standard molar enthalpy change of evaporation is plotted (on the y-axis) against the boiling temperature (on the x-axis) most of the points lie on or close to a straight line whose gradient (or slope) is $\triangle H^{\theta}_{evap}/T_{b}$ (Fig. T.6).

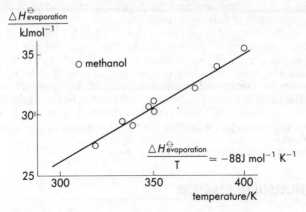

Fig. T.6

$\triangle H^{\theta}_{evap}/T_{b} = \triangle S^{\theta}_{evap}$ is the standard molar entropy change when a substance changes from the liquid to the gaseous state at its boiling point.

A wide range of liquids obey Trouton's rule because the change in volume when one mole of liquid forms a gas is similar for most liquids and the **molar gas volume** is similar for most gases. Consequently, the increase in the **entropy** (the dispersion of matter and energy) is about the same for one mole of most liquids. However, liquids with **hydrogen bonding** between molecules have $\triangle S^{\theta}_{evap}$ values that are greater than Trouton's rule would predict. For example, $\triangle S^{\theta}_{evap}$ [methanol] is 104 J K^{-1} mol^{-1}. The hydrogen bonding in water and the **alcohols** gives these liquids a structure and makes their molecules more ordered than most liquids so the **entropy change** on evaporation is greater than that for most liquids.

ULTRAVIOLET ABSORPTION SPECTROSCOPY

Ultraviolet absorption spectroscopy is an analytical technique for determining the molecular structure of a substance by measuring the wavelengths of ultraviolet radiation absorbed by the substance. Ultraviolet radiation emitted by a hydrogen discharge lamp is scanned, selected and focused, by a system of mirrors and a prism or reflection grating through the sample cell, on to a photoelectric or photoconductive detector. The sample cell and windows are made of quartz glass. The sample is usually dissolved in a solvent, such as an alcohol, heptane, trichloromethane or water, which is transparent to ultraviolet light. The output from the detector is automatically plotted against the wavenumber of radiation transmitted.

When the energy of the ultraviolet radiation matches the transitions in the energies, particularly of the bonding electrons within the molecules of the sample, absorption of the radiation occurs. UV absorption spectra are important in organic chemistry because they provide information about the structure of arenes and unsaturated compounds, especially when the molecules contain a conjugated system of alternating single and double bonds in their structure. UV spectroscopy can be used to detect, identify and measure the amount of a wide variety of inorganic as well as organic compounds.

◀ Infra-red absorption spectroscopy ▶

UNIMOLECULAR REACTION

A unimolecular reaction is a step in a chemical reaction that involves one species. If the rate of reaction which proceeds by a series of steps is governed by a slow, rate-determining unimolecular step, the overall chemical reaction is sometimes called a unimolecular reaction. Howeve, it is better to restrict the use of the terms unimolecular and bimolecular to the individual steps in a reaction. For example, the rate of hydrolysis of tertiary halogenoalkanes is governed by the unimolecular reaction step in which one molecule of the halogenoalkane undergoes heterolytic fission:

$$(CH_3)_3C-Br \rightarrow (CH_3)_3C^+ + Br^-$$

The molecularity of a reaction must not be confused with the order of a reaction. The terms unimolecular reaction and first-order reaction have different meanings.

UNIT CELL

Unit cells are the smallest identical blocks of **atoms, ions** or **molecules** which can be stacked together to fill space completely and to reproduce the regular arrangement of the atoms, ions or molecules in the **crystal structure**. Only seven different shapes of block are possible if the stacking together of identical unit cells is to fill space completely. These three-dimensional geometrical shapes are defined by the lengths (A, B, C) of its sides and the angles (x, y, z) between the sides – see Fig. U.1.

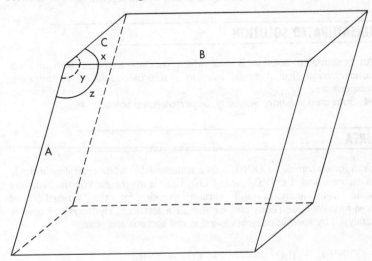

Fig. U.1 Crystal systems

The seven possible shapes for the unit cell determine the seven major crystal systems: cubic, tetragonal, orthorhombic, rhombohedral, hexagonal, monoclinic and triclinic. The dimensions and composition of the unit cell of a **crystal lattice** can be determined by X-ray **diffraction**.

UNSATURATED HYDROCARBON

An unsaturated hydrocarbon is a compound, of carbon and hydrogen only, in which at least two atoms share four or six **valence electrons**, so that the molecule contains at least one **double** or **triple bond**.

Three broad classes of unsaturated hydrocarbons are **alkenes**, **cycloalkenes** and **alkynes**. Cycloalkanes may be classed as unsaturated in so far as their general formula is the same as that of the alkenes, C_nH_{2n}, but the molecular structure of cyclohexane contains only single covalent bonds, Benzene, C_6H_6, is unsaturated in so far as its **empirical formula** is the same as that of ethyne, C_2H_2, and its ring structure is sometimes represented as though it contains single and double bonds. However, although benzene can undergo some of the addition reactions expected of unsaturated hydrocarbons, most of the reactions of benzene and its derivatives are substitution reactions expected of saturated hydrocarbons.

UNSATURATED SOLUTION

An unsaturated solution is one which contains a solute with a lower **concentration** than it would have in a **saturated solution** at the same temperature.

◀ Saturated solution, Solubility, Supersaturated solution ▶

UREA

Urea (or carbamide), $CO(NH_2)_2$, is a water-soluble white crystalline diamide of carbonic acid, $CO(OH)_2$ (or H_2CO_3). Urea is synthesised on the industrial scale from ammonia and carbon dioxide for the production of urea-formaldehyde resins and for use as a fertiliser. **Hydrolysis** of urea is catalysed by specific **enzymes** found in soil bacteria and plants:

$$CO(NH_2)_2 + H_2O \xrightarrow{\text{enzymes}} CO_2 + 2NH_3$$

The resulting ammonia molecules are oxidised to nitrate ions which the plants can absorb. Urea is the main way by which mammals excrete nitrogen.

VALENCE ELECTRON

The valence electrons of an **atom** are the outer energy shell electrons which are transferred or shared during chemical bonding.

VALENCE SHELL ELECTRON PAIR REPULSION THEORY

According to the valence shell electron pair repulsion (VSEPR) theory, the shape of a simple molecule may be predicted and explained in terms of the repulsions between pairs of **valence electrons** in its atoms. The theory does not give a quantitative prediction of **bond angles**, but the shape of a simple molecule can be predicted well enough to decide if any **polar bonds** will make the molecule polar with a permanent **dipole moment**. There are three steps to follow to predict molecular shape.

Step 1

The displayed formula is written showing all bonded electron pairs by dashes (−) and all non-bonded pairs by two dots (:)

$$\begin{array}{ccccc}
\text{Cl} & \text{H} & \text{H} & \text{H} & \text{:O:} \\
| & | & | & | & || \\
\text{B−Cl} & \text{H−C−H} & \text{H−N:} & \text{:O:} & \text{C} \\
| & | & | & | & || \\
\text{Cl} & \text{H} & \text{H} & \text{H} & \text{:O:}
\end{array}$$

Step 2

The bonded and non-bonded electron pairs are assumed to move equally as far apart as possible from each other. A double-bonded pair (=) is treated as a single-bonded pair.

Cl | B−Cl | Cl	H | C H H H	Ṅ H H H	Ö H H	:O: || C || :O:
trigonal 120°	tetrahedral 109.5°	pyramidal	non-linear	linear 180°

Step 3

The rule that *repulsion between a non-bonded and a bonded pair is greater than that between two bonded pairs and two non-bonded pairs repel each other even more strongly* is followed to adjust any bond angles affected by it. For example, the H–N–H and H–O–H bond angles in ammonia and water respectively are 'squeezed together' by the stronger repulsions from non-bonded electron pairs to less than 109.5° (tetrahedral angle).

pyramidal
107°

non-linear
105°

VAN DER WAALS' EQUATION

$$(P + a/V_m^2)(V_m - b) = RT$$

This is an equation of state for real gases. It was proposed in 1873 by the Dutch chemist, J.C. van der Waals. P is the pressure and V_m is the volume of one mole of gas; a is a constant to allow for the reduction in ideal gas pressure caused by the attractive forces between the molecules; b is the effective volume in one mole of gas of the molecules themselves, so $(V_m - b)$ is the ideal gas volume.

◀ Ideal gas equation ▶

VAN DER WAALS' FORCES

Van der Waals forces are weak short-range forces between atoms and molecules arising from the attraction between **dipoles**. The term is usually applied to the very weak short-range forces operating between **atoms** in all substances. In this case, these forces arise from the movement of the **electrons** in relation to the **nuclei** producing weak instantaneous dipoles that attract one another. The more electrons there are in an atom, the more instantaneous **dipole–dipole attractions** there will be. Van der Waals forces hold together the atoms when a **noble gas** is liquefied. These forces also account for the increase in boiling point with increasing **molar mass** of the noble gases, the **halogens** and members of **homologous series** such as the **alkanes**. The very short-range nature of van der Waals forces explains the decrease in boiling point with increasing branching of isomeric alkanes (see Fig. V.1).

Van der Waals forces is a term sometimes used to embrace the attractions between permanent dipoles and between a permanent dipole and an induced dipole. Van der Waals forces between permanently polar molecules are then referred to as dipole–dipole attractions. Attractions involving induced dipoles are known as dispersion or London forces.

◀ Intermolecular forces ▶

Isomers of C_6H_{14}	B.pt/°C
hexane	68
3-methylpentane	63
2-methylpentane	60
2,3-dimethylbutane	58
2,2-dimethylbutane	50

Fig. V.1 Boiling temperatures (°C) of isomers

VAPOUR PRESSURE

The vapour pressure of a liquid or solid substance is the pressure of the substance in the gaseous state in **dynamic equilibrium** with the liquid or solid in a closed container. Vapour pressure depends only on the nature of the substance and the temperature. It does not depend on the amount of liquid or solid substance present in the container. If a liquid is put into a barometer tube containing only mercury, the vapour pressure of the liquid causes the height of the mercury column to drop, and the space above the mercury becomes saturated with the vapour (see Fig. V.2). The more volatile the liquid, the greater the drop and the greater the saturated vapour pressure of the liquid. If the temperature is increased, the height of the mercury column falls even further, because the vapour pressure increases as more of the liquid evaporates (Fig. V.3).

Barometer tubes filled with mercury then inverted into trough of mercury
Different liquids injected into the space above the mercury in the tube.

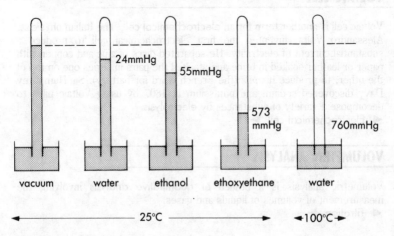

vacuum	water	ethanol	ethoxyethane	water

24mmHg
55mmHg
573 mmHg
760mmHg

25°C — 100°C

Fig. V.2 Vapour pressure of liquids

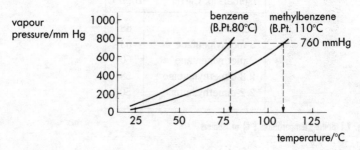

Fig. V.3 Change of vapour pressure with temperature

Vapour pressure and boiling point

If a liquid in a closed container is heated, the vapour pressure will increase and eventually exceed the pressure of the atmosphere outside the container. If the container is not closed, the liquid will boil when its vapour pressure reaches atmospheric pressure. The temperature at which the saturated vapour pressure equals the pressure of the atmosphere is the **boiling point** of the liquid. The vapour pressure of a liquid at a given temperature is lowered when the liquid acts as a solvent for another substance. And the boiling point of a liquid at atmospheric pressure is raised when the liquid acts as a solvent for another substance. The **lowering of the vapour pressure** and the **elevation of the boiling point** of a liquid are **colligative properties**.
◀ Raoult's law ▶

VOLTAIC CELL

Voltaic cell is another term for an **electrochemical cell**. The Italian physicist, Alessandro Volta, invented the first electrochemical cell to produce a substantial current of electricity. He separated discs of zinc and copper with paper or leather soaked in brine and stacked the pairs in series one on top of the other, to produce his *pile* (the French word for battery). Sir Humphrey Davy discovered sodium and potassium in 1807 by using Voltaic piles to decompose a variety of substances by **electrolysis**.
◀ Electrochemical cell ▶

VOLUMETRIC ANALYSIS

Volumetric analysis is a branch of **quantitative analysis** involving the measurement of volumes of liquids and gases.
◀ Titration ▶

WATER OF CRYSTALLISATION

Water of crystallisation is the water present in definite amounts in hydrated crystals. Many ionic compounds crystallise from **aqueous** solution as **hydrates**. The water of crystallisation is often part of a hydrated cation. For example, in copper(II) sulphate-5-water, four of the water **molecules** are attached to the central copper ion by **dative bonds** to form a square planar **complex** cation.

◄ Hydrate ►

WEAK ACID

Weak acids are **proton** donors which only partially ionise in water. Ethanoic acid is a typical weak acid. At 25°C its **acid dissociation constant**,
$K_a = 1.7 \times 10^{-5}$ mol dm^{-3}
and its pK_a = 4.8. These values are typical for **carboxylic acids**. Fewer than 10 per cent of the ethanoic acid molecules ionise into hydrogen ions and ethanoate ions:

$$CH_3CO_2H(aq) + H_2O(l) = H_3O^+(aq) + CH_3CO_2{}^-(aq)$$

Hydrofluoric acid (pK_a = 3.3) in contrast to the other aqueous hydrogen halides, is a weak acid because **hydrogen bonding** causes its molecules to dimerise and this decreases the number of protons donated to the water molecules:

$$H_2O(aq) + H-F\cdots H-F(aq) \rightarrow H_3O^+(aq) + [F\cdot\cdot H-F]^-(aq)$$
$$\text{dimer}$$

Other weak acids include carbonic acid (H_2CO_3(aq), pK_a = 6.4), nitrous acid (HNO$_2$(aq), pK_a = 3.3) and sulphurous acid (H_2SO_3(aq), pK_a = 1.8).

WEAK BASE

Weak bases are proton acceptors which only partially ionise in water. Ammonia is a typical weak base. At 25°C its **base dissociation constant**, $K_b = 1.8 \times 10^{-5}$ mol dm^{-3} and its pK_b = 4.8. Fewer than 10 per cent of the ammonia molecules ionise into ammonium ions and hydroxide ions:

$$NH_3(aq) + H_2O(l) \rightleftharpoons NH_4^+(aq) + OH^-(aq)$$

Other weak bases include the **amines**, which are derivatives of ammonia. Phenylamine ($pK_b = 9.4$) is a weaker base than ammonia because the **lone pair** of electrons on the nitrogen atom participate in the delocalised π-electron system of the benzene ring. This stabilises the free base molecule and makes the lone electron pair less readily available for **dative bonding** with a proton.

X-RAY

X-rays are penetrating high-energy electromagnetic radiations emitted from an element, usually a metal, when it is bombarded by high-energy electrons.

X-RAY CRYSTALLOGRAPHY

X-ray crystallography is the technique of determining crystal structures by X-ray diffraction and the analysis of X-ray diffraction patterns. The wavelengths of X-rays are similar in size to the interatomic distances in solids. So when a monochromatic beam of X-rays is passed through crystalline solids, the repeated planes of atoms cause the crystals to act as gratings and produce diffraction patterns. A single crystal produces a pattern of spots of varying intensities, whereas a powdered crystal produces a series of concentric rings of varying intensities. From the relative positions and intensities of the spots or rings it is possible, with the aid of a computer, to determine the dimensions and composition of the unit cell of the crystal. And in the case of molecular crystals, the structure of the molecules may be determined.

The extent to which X-rays are diffracted depends, among other things, upon the number of electrons in the atoms of the crystal. Consequently, hydrogen atoms are the most difficult to detect by X-ray crystallography and are usually located by other techniques such as neutron diffraction and nuclear magnetic resonance spectroscopy.

X-RAY DIFFRACTION

X-ray diffraction is the scattering of X-rays by the planes of atoms in a crystalline solid acting as a grating. In 1912 Sir Lawrence Bragg derived his law relating the wavelengths (λ) of the X-rays and the angles (θ) at which their diffractions occur to the spacing (d) between the crystal lattice planes:

$\lambda = 2d\sin\theta$.

This basic equation of X-ray diffraction is known as the *Bragg equation*.

YEAST

Yeast is the name for a group of unicellular organisms, some of which are used in baking, brewing and wine-making. Yeasts known as saccharomycetaceae produce **enzymes** which catalyse the breakdown of starches and sugars aerobically into carbon dioxide and water or anaerobically into carbon dioxide and ethanol.

ZERO-ORDER REACTION

A zero-order reaction is a reaction with a differential **rate equation** of the form $-d[R]/dt = k$, an integrated rate equation of the form $[R] = -kt + [R]_i$ and a **half-life** that depends upon the concentration. $[R]_i$ and $[R]$ represent the **concentration** of a reactant R initially (at $t = 0$) and at time t, respectively; k is the **rate constant**; $-d[R]/dt$, the rate of decrease of the concentration of a reactant with time, represents the rate of reaction which, in a zero-order reaction, is directly proportional to the concentration raised to the *zero*th power. In other words, the rate is independent of the concentration of the reactant – see Fig. Z.1.

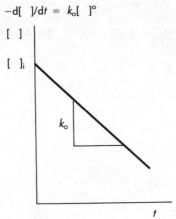

$$-d[\ ]/dt = k_o[\ ]^0$$

Fig. Z.1 Zero-order reaction kinetics

The tungsten-catalysed decomposition of ammonia into nitrogen and hydrogen is a zero-order reaction because the rate of decomposition of the ammonia is independent of the concentration of $NH_3(g)$. In a pseudo-**first order** reaction such as the **hydrolysis** of sucrose to fructose and glucose, the order of reaction with respect to water appears to be zero because any change in the concentration of the water is negligible. The reaction of iodine with propanone in aqueous acid is zero-order with respect to iodine because iodine molecules are involved only in those fast reaction steps which follow the **rate-determining steps** and which are controlled by the concentration of propanone and of hydrogen ion.

ZIEGLER–NATTA CATALYSTS

Ziegler–Natta catalysts are titanium(IV) chloride and an aluminium trialkyl used in the production of high-density stereoregular addition polymers. Using Ziegler–Natta catalysts, the **polymerisation of alkenes** can be carried out at temperatures below 100°C and at atmospheric pressure to produce **isotactic straight-chain** poly(alkenes) very efficiently. Strictly speaking, the aluminium trialkyl is not a catalyst because it is consumed in the reaction.

The process was originated by the German chemist Karl Ziegler in 1953 and further developed by the Italian chemist Guilio Natta in 1954. They were awarded the Nobel Prize in 1963.

ZWITTERION

Zwitterion is the term for an ion which has a positive and a negative charge on the same molecular structure. **Amino acids** and other compounds which have an acidic and a basic group in the molecule tend to exist as zwitterions; for example, glycine (α-aminoethanoic acid) forms a crystalline solid with an unexpectedly high melting point because the crystal is composed of zwitterions $^+NH_3CH_2CO_2^-$.